Pentre Farm, Flint 1976 – 81

An official building in the Roman lead mining district

T. J. O'Leary
with
Kevin Blockley and
Chris Musson

BAR British Series 207
1989

B.A.R.

5, Centremead, Osney Mead, Oxford OX2 0DQ, England.

BAR 207, 1989: 'Pentre Farm, Flint, 1976-1981'

ISBN 9780860546344 paperback
ISBN 9781407318301 e-book
DOI https://doi.org/10.30861/9780860546344
A catalogue record for this book is available from the British Library
This book is available at www.barpublishing.com

Contents

LIST OF PLATES

LIST OF FIGURES

LIST OF TABLES

ACKNOWLEDGEMENTS

The excavations at Pentre Farm were undertaken by the Clwyd-Powys Archaeological Trust in two major seasons, the first in 1976/77, the later in 1981 (Fig 6).

Funding for the excavations was provided by Cadw: Welsh Historic Monuments, who also financed the post-excavation work.

The excavations were directed by T J O'Leary, who also prepared the excavation report and discussion section. Mr C R Musson (then director of the Clwyd-Powys Archaeological Trust) undertook extensive editing of the report in its initial stages. Final editing and organisation of the report for publication was undertaken by myself.

Illustrations were prepared by Brian Williams (structural); Peter and Janet Webster (pottery and small finds) and Kathy Holland (small finds and final corrections).

Thanks go to Professor Barri Jones for helpful comments during the excavations and to Sheila Gibson for comments on inital drafts of the reconstruction drawings.

I should like to thank Mike Yates (Cadw) and Jeff Davies (University College, Aberystwyth) for comments on a final draft of the report, and Jenny Britnell and Chris Martin for transforming a much amended text to a state worthy of publication.

Thanks also go to Bill Britnell (director of CPAT) for constant help and encouragement throughout the final stages of publication.

I am indebted to all those who prepared specialist reports for this volume (noted in the contents list).

A special note of thanks go to all those who helped in all aspects of excavation and post-excavation work, particularly the inhabitants of Flint, who helped the work proceed smoothly.

Kevin Blockley
Llidiartywaen, February 1989.

CONVENTIONS USED ON PLANS

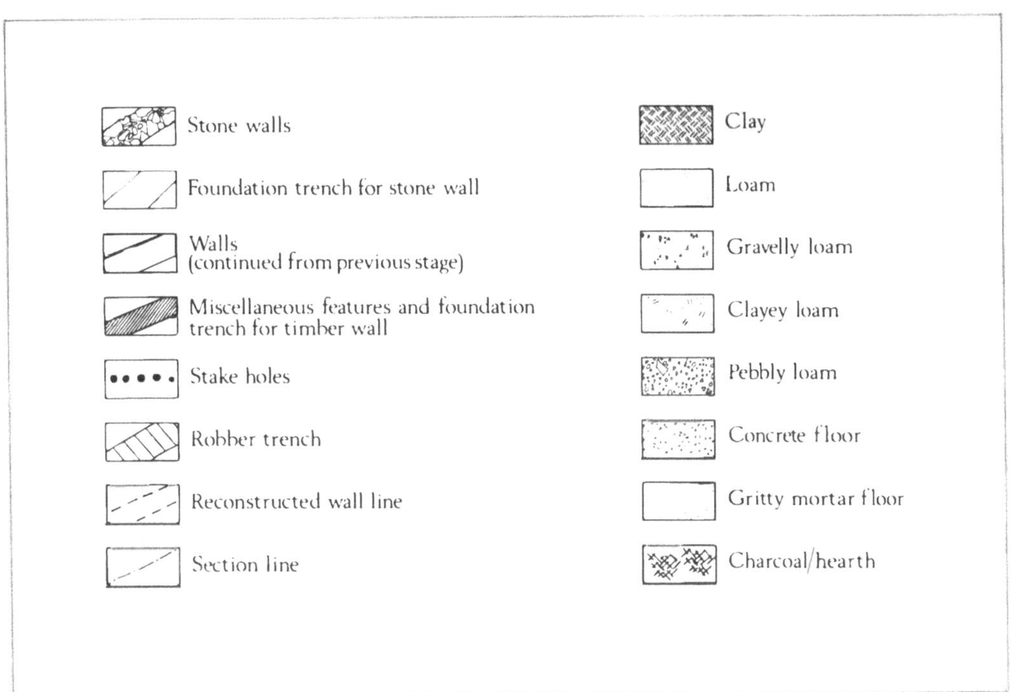

Figure 1

Introduction (T J O'L)

PHYSICAL ENVIRONMENT

The site of Pentre Farm is situated in the hamlet of Oakenholt, 1 km east-south-east of the town of Flint, at SJ 254 723. It lies astride the 10 m contour line on land sloping gently towards the marshes on the edge of the Dee estuary (Fig 2). The site was hemmed in by the tracks of the main east-west railway on the north; the main east-west road (A548) on the south; a bungalow to the west and the buildings of Pentre Farm to the east

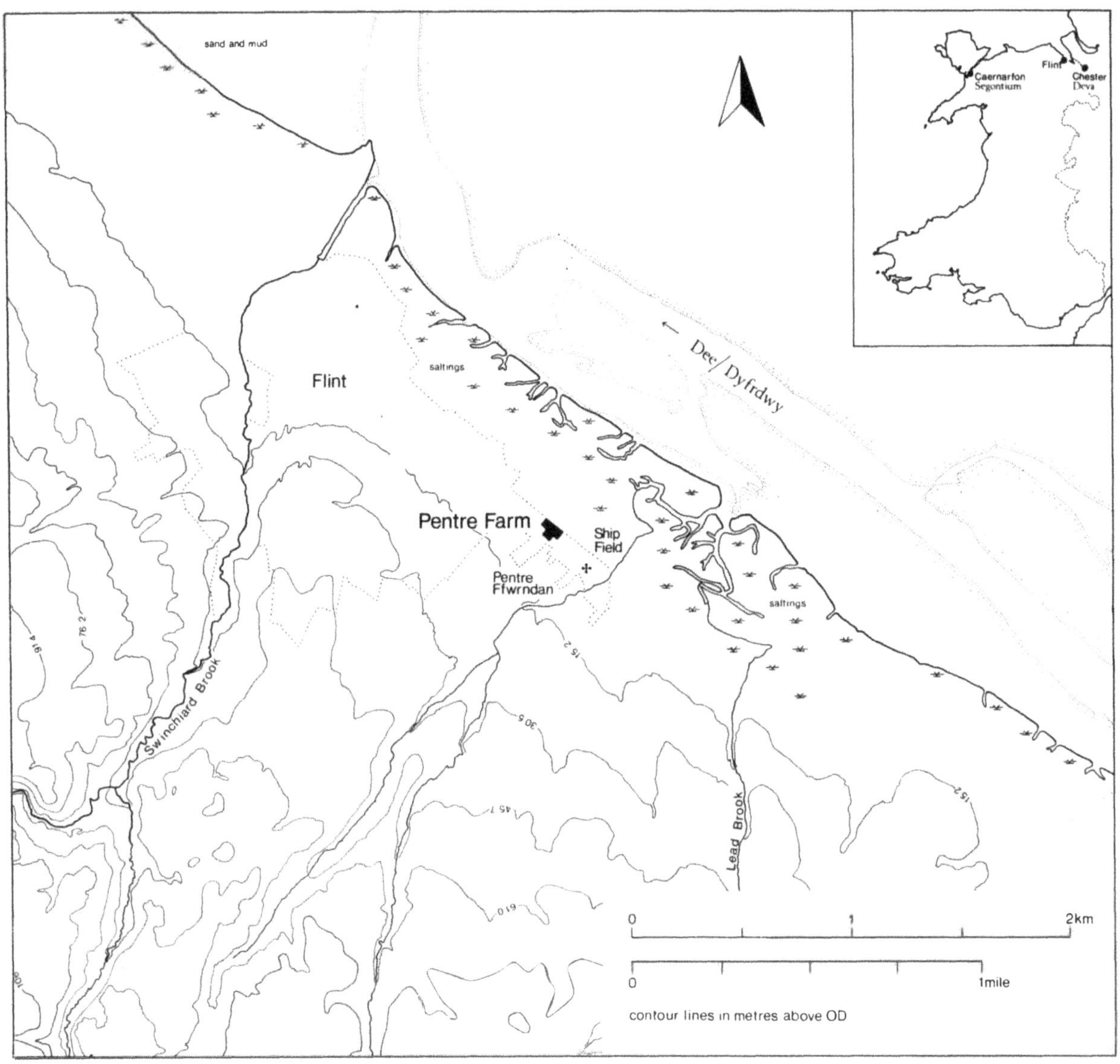

Figure 2: Location plan of Pentre Farm, Flint

(Fig 3). The region around Pentre Farm is most easily described in terms of the old county of Flintshire which consisted, for the most part, of a long plateau of fairly high ground (less than 310 m), encircled, except in the south and the south-west, by a margin of lower land skirting the sea and the estuary of the Dee, and extending along the valleys of the Dee, Alun & Clwyd.

The plateau forming the centre of Flintshire is formed by carboniferous limestone lying above the Silurian rocks. Around Halkyn, just to the south of Flint, the limestone contains lead ore in workable quantities. On top of the limestone is a series of sandstones known as the Millstone grit series. In the north of Flintshire the series is represented by the Holywell shales, from which sandy beds are almost absent. To the east and lying on the limestone are the Coal Measures of the Flintshire coalfields.

These consist mainly of shales with subordinate sandstones and a number of coal seams. These rocks are less resistant to weathering than older rocks and therefore they now form the low-lying ground of the country. To the east of Flint Red Sandstone of the Triassic System overlies the coal measures but is largely hidden by a cover of glacial Boulder Clay, covered along the Dee estuary by river alluvium (Davies 1949).

Today the coastal strip on which Flint lies, between Fynmor-Groew and Connah's Quay, is an almost continuous built-up area with much industrial property partially derelict. The river Dee is badly silted owing to the building of sea walls, the earliest of which was a causeway perhaps built by Earl Hugh of Chester. The walls prevent scour and the steady widening of the estuary seawards also prevents the ebb tide from developing a strong scouring effect. In the Roman period, however, shoal conditions were different and sea level was possibly higher, (the River Dee is marked on Ptolemy's map of Britain as *Seteia Est)*. Then too, the industrial and denuded landscape of present day Flint would have been covered by forest, like one south-east of Holywell, referred to in the Domesday Book.

HISTORY OF DISCOVERY

The first reference to historical remains in the area was made in 1778 by Thomas Pennant (1778) who remarked on both a local tradition that a large town had, in the past, stood at Croes Atti and on evidence of building foundations turning up under the plough. He also thought notable the occurrence of tons of lead ore and waste from various places here and in Northop; and in particular at Pentre Ffwrndan (furnace village), to the south of Pentre Farm, a name suggestive to him of the antiquity of lead smelting in this area. He published, too, descriptions and drawings of many coins, brooches and other objects of second to early third century date. His conclusion was that the Romans extracted and exported lead, having a settlement in or around the town of Flint.

The six-inch Ordanance Survey map of 1900 (sheet ix NE) marks the site of an 'Ancient Furnace' discovered in 1840 under the south-west part of the Mill Dam belonging to Croes Atti Mill, about 200 m south of St David's Church.

The next major group of finds were reported by Judge Wynne Ffoulkes (1856), who noted discoveries made in Ship Field, to the south-east of the Chester Road, opposite St David's church. Five skeletons were found in crude stone cists, which were not intact, and close by, fragments of *terra sigillata* were present as well as sherds of mortaria and amphorae, tile fragments, animal bones and two coins, one of which was believed to be Trajanic. In the north-west of Ship Field was a 54 ft length of stone wall, bonded with a stiff clay and with easterly returns at each end. Nearby there was a hole sunk to the 'rammel' (*c* 1.75 m deep) 'with the same rude flagging'. What was probably a well or a tank was also discovered. Finally, Ffoulkes remarks on some coins, supposedly 'Caesar's' which were said by an old man to have been discovered by his father whilst digging some foundations just to the south-east of the brook which had once formed the southern boundary of the Ship Field.

Henry Taylor, writing his *Historic Notices of Flint* (1883) noted discoveries of pieces of glass, tiles, and other evidence of Roman occupation, found at Croes Atti. Taylor also referred to a coin of Commodus (AD 177-92) discovered when trenching Chester Road opposite St David's Church. Forty years later but still prior to any excavation in the area the existing evidence was summarised: 'All that can now be said is that the place was probably never more than a small mining village whither the lead was brought down from the hills, where it was smelted and shipped to Chester. Lead abounds in the Halkyn Hills a mile or two to the south; an outcrop of cannel coal occurs close to Pentre and the Romans used cannel and other coal for smelting ... The stream would be useful for smelting purposes and its mouth provided a little harbour. There may also have been an easy ford and way across the sands here to the Wirral Peninsula and hence to Chester; such a route was used by Edward I to transport his armies across the estuary. Pentre was occupied in the second and third centuries AD, if not earlier and apparently by a population with Celtic tastes; possibly the industry and population dwindled by the fourth century' (Taylor 1923).

In the autumn of 1923 and 1924 excavations were carried out over a total of four weeks in Ship Field (*c* 200 m south east of Pentre Farm), and the results of these are crucial to any understanding of the Pentre Ffwrndan Roman settlement (Atkinson and Taylor 1924, parts i & ii). In 1923, several narrow trenches up to 60 m long, revealed remains of two furnaces, one of which (Furnace B) seemed to the excavators to represent the remains of a lead-smelting furnace, having associated with it ore, lumps of slag, fragments of lead, charcoal, and stones coated with vitreous slag. There were also shed walls, a flagged area, an occupied area, and a shallow pit (Davies 1949).

In 1924, eleven more trenches were cut in which the remains of up to six or seven furnaces together with 'shed walls' and two roughly paved areas were uncovered. While not all of the furnaces bore signs of association with lead working, one (Furnace L) was described as a 'group of stones set in clay and covering an area of 4 by 4 ft. Several of the stones were highly glazed, evidently by intense heat, while at two places, run lead and lead slag were firmly embedded among the stones' (Davies 1949). Pottery dated most of the structures to the first half of the second century.

No further work was carried out in the area until 1933-4, when Petch excavated opposite Ship Field. He uncovered remains of a Roman building constructed of sandstone, the eastern end of which had been destroyed by the Chester Road. To the south-east was part of another room beyond a paved area. Within the footings of the first building were two sherds of Hadrianic *terra sigillata* (Petch & Taylor 1925). Petch felt that he had not shown the existance of extensive lead smelting here, in spite of the suitability of the location and the easy access to lead supplies on Halkyn

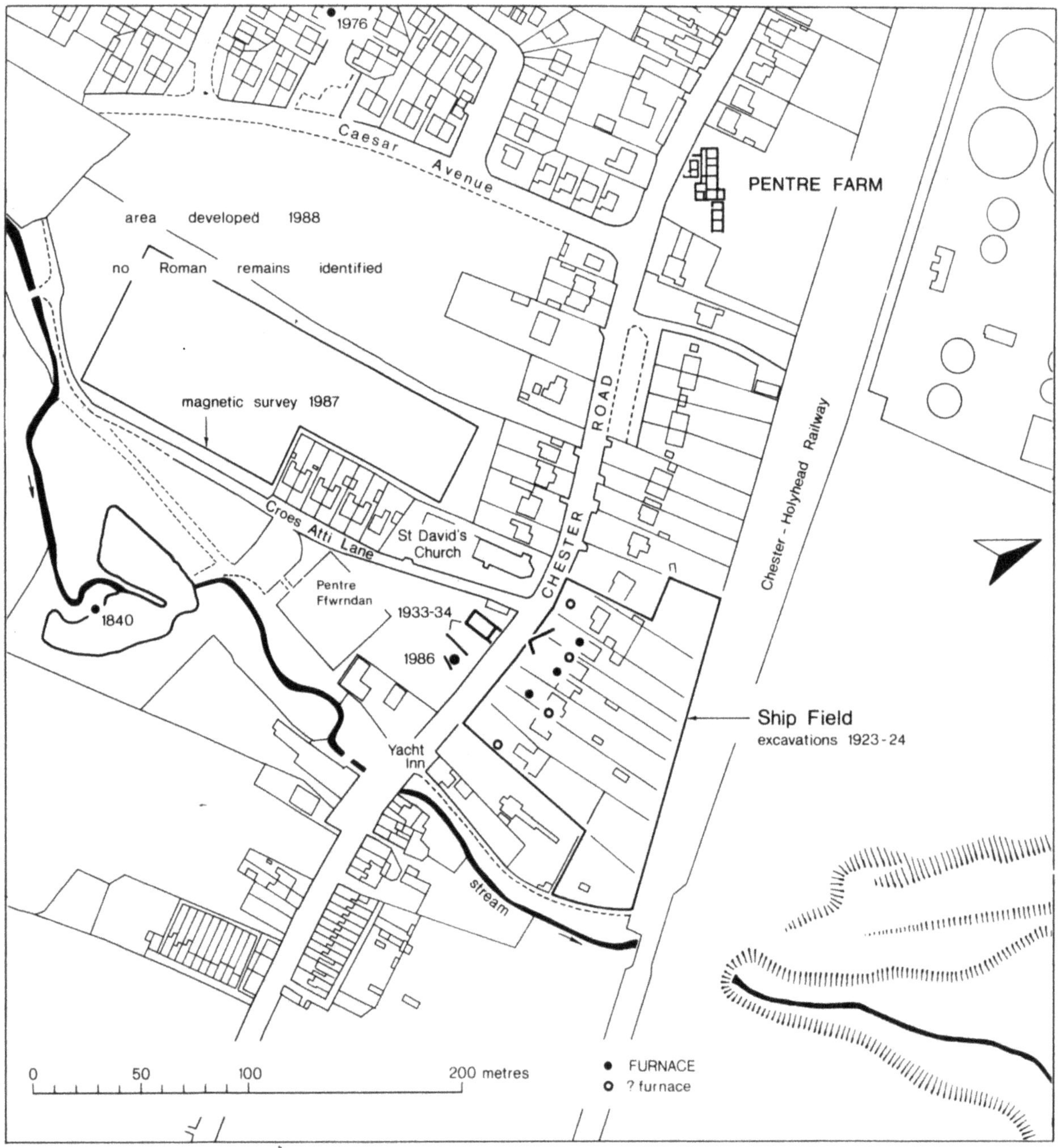

Figure 3: Romano-British sites in Flint, scale 1:200

mountain. The only other excavation work prior to the present excavations was carried out by Hayes in 1958 and 1961. The first of these, in the south of Ship Field, revealed a shallow pit and the second penetrated the hypocaust system of the bath-house more fully investigated in the present excavations. Hayes wrote: 'This new Roman building leaves a gap of about 230 metres between it and the nearest remains found in the earlier excavations. This limited excavation suggests, however, that it has a different character from the building found on the rest of the site and may be the residence of some important executive concerned with the lead smelting' (Hayes 1961). As a result of this discovery the site was scheduled as an Ancient Monument (No 122 in the Flintshire list) in September 1961. When, in early 1976, proposals were made to build on the site, investigation of the remains was considered necessary. At the invitation of the Clwyd-Powys Archaeological Trust an exploratory excavation was carried out by Professor G D B Jones of Manchester University in June 1976. This provided positive evidence of Roman occupation in the late first or second centuries AD, and led to the more extensive project now reported in this monograph.

[Professor Jones also noted at this time a further furnace to the south-east of Pentre Farm during building works (*Britannia*, 8 (1977), 358).

Trial excavations on a plot of derelict land at the junction of Chester Road (A458) and Croes Atti Lane were conducted by the Trust in 1986, to evaluate the site prior to the proposed development of an amenity area (Weetman 1986, 15-16).

Further traces of masonry and timber buildings were found, to the south of those identified by Petch, including a small furnace and lead slag and ore (ibid, 15).

A magnetic survey was carried out in 1987 on behalf of the Trust by Dr W T C Sowerbutts of the Department of Geology, University of Manchester, within the area of a proposed housing development to the south of Pentre Farm (Fig 3). This produced anomalies suggesting additional furnaces, but these were not confirmed by subsequent trial excavation, and no additional Roman remains were identified in this area during the course of development in 1988. KB]

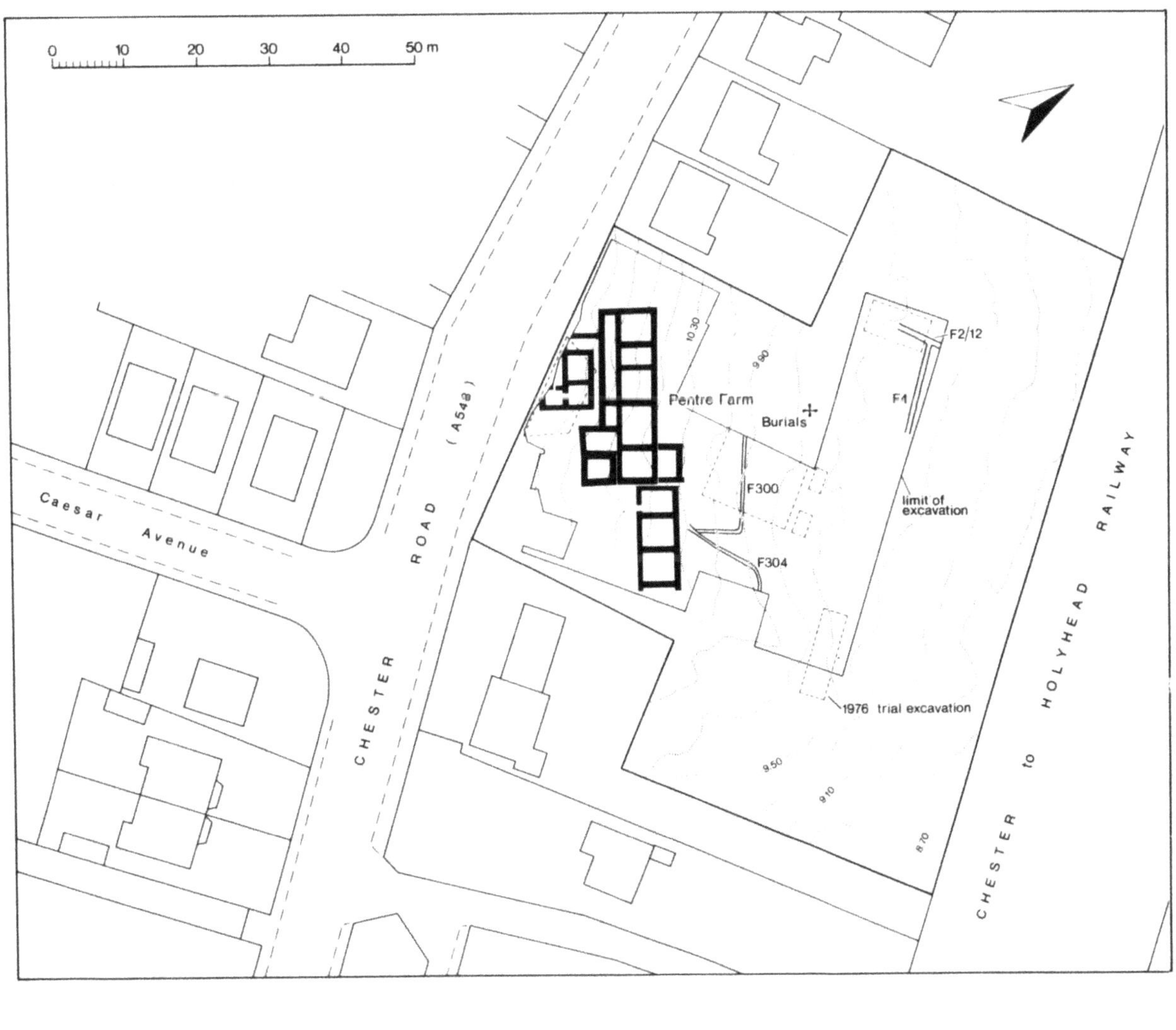

Figure 4: Location plan of excavation trenches, scale 1:100

Summary (KB)

Stage I

The initial phase of the complex is represented by three ranges of buildings of clay and timber construction built soon after AD 120. A yard, incorporating a D-shaped ornamental pool, lay in the heart of the complex. It is possible that an eastern range may have been built (outside the limit of excavation) to produce a symmetrical layout. The complex is thought to represent an official residence associated with lead mining in the district.

Stage II

Few changes were undertaken save new rooms being added to the West Range.

Stage III

In the Antonine period the verandah and external corridors were removed from the South Range, part of which was used for lead working, and of the West Range demolished.

Stage IV

This stage saw considerable rebuilding of the complex sometime in the late second century. The North Range and part of the West range were demolished and a masonry building (North-west Range) constructed, a corridor running the length of the building.

Stage V

An additional stone building (North Range) was built and the remaining section of the West Range replaced by a heated room. The timber South Range remained standing.

Stage VI

Some minor alterations were undertaken, in the late second to early third centuries, on the existing buildings. The heated room was replaced by a bath-house, comprising four main rooms, one containing a plunge bath.

Stage VII

In the early to mid third century the complex saw the demolition of the South Range and addition of three rooms (one timber, two masonry) on the south side of the North-west Range. The central room may have been a tower (Fig 16) providing an imposing front elevation.

Stage VIII

The majority of the complex was robbed of its stone in the mid third century, although additional floors and walls in some of the southern rooms indicate a continuation of occupation.

Figure 5: Development plans of the complex, scale 1:600

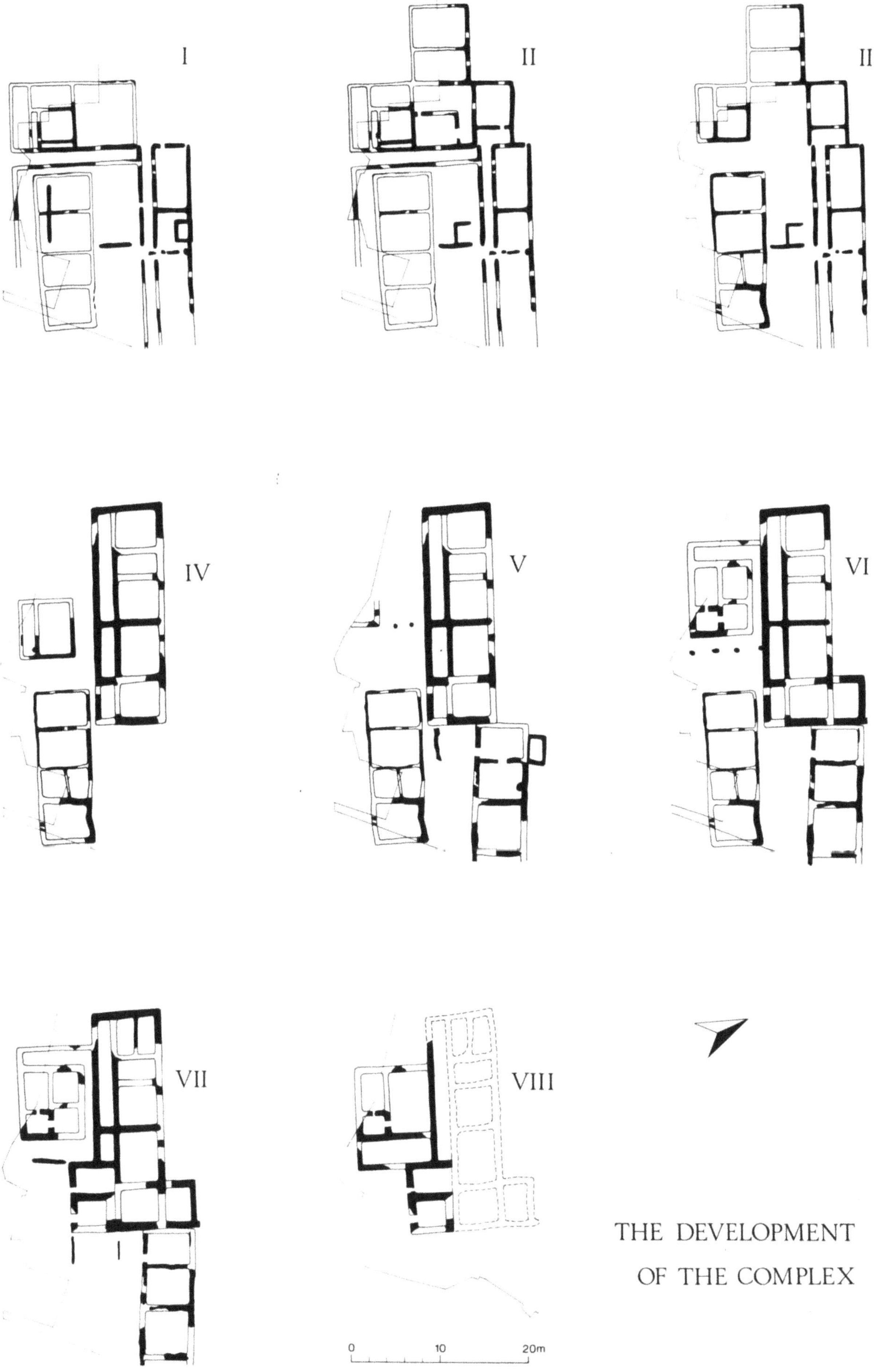

THE DEVELOPMENT
OF THE COMPLEX

Part 1: The site (TJO'L)

PRE-COMPLEX OCCUPATION

Prehistoric

During the excavation, nine pieces of undatable, glacial or beach flint were recovered. They are described in the archive report.

Roman

Running across the western part of the site, from south-east to north-west was a ditch F0 (Fig 18). Its course was traced in plan in the north-west and in the south-east by recording the point where the stratification in the complex had subsided into the ditch as its fill settled. It was excavated only in two *sondages* (Fig 18).

The northern *sondage* (C on Fig 18) revealed that the ditch at this point was flat-bottomed, 0.80 m wide, with sides sloping outwards to make a total width of 1.20 m at the top. Most of the ditch was filled by a fawn homogenous clay with small loam inclusions, either the weathering of natural subsoils or, more probably, the result of deliberate back filling with clay upcast from ditch F500. The clay was covered by a charcoal-laden clay-loam associated with the destruction levels of Stage VII or VIII.

Where the section of the ditch was revealed in the side of *sondage* D, the profile, slightly distorted in the oblique cut, was a flattened V-shape with a slightly rounded base. The lowest fill of rubble was covered by a fine fawn silt which rested on a possible re-cut, and beneath on a dense orange-brown clay containing a lens of green clay.

Sediment analysis of the lowest ditch-fill confirmed that this had not been water deposited. Dr Shackley suggests that it might have been formed by the crumbling of the sides of the ditch but it is equally possible that it resulted from deliberate backfilling. This is implied by the nature of the fills, rubble at the bottom, and their horizontal rather than curved profile. The silt above the rubble met the sides of the ditch at the points where there were irregularities in their slopes and it is possible that the layer represents an accumulation in the ditch after it had been re-cut. The topmost ditch-fill at the point of Section C was cut by a Stage 1 foundation trench, may have been a deliberate levelling of the ditch before the construction of the complex.

Ditch FO is the only feature certainly pre-dating the Pentre Farm complex. The several sherds of pottery found in it included material dated typologically to the period AD 90-120. A possibly intrusive sherd from the ditch-fill, might support an early-Hadrianic starting date for the complex, a date otherwise based on the black-brownish ware built into pool F27, a feature possibly of Stage I (p 12).

[A more detailed assessment of the distribution and function of pre-complex features is not possible given the disturbance by later features, and only partial excavation of the ditches. KB]

STAGE I

The evidence for Stage I (Fig 6) rested on a series of foundation trenches for clay and timber walls forming three ranges on the northern, southern and western sides of a subdivided yard, any eastward continuation of the complex lying beneath modern access paths, yards and buildings. Much of the Stage I stratification, including some of the foundation trenches, had been disturbed or truncated by later phases of development. In other parts fragmentary surfaces and other layers were removed without examination during salvage work towards the end of excavation aimed at uncovering the whole of the wall layout.

Allowing for minor variations mentioned in the text, the general method of construction was as follows. The foundation trenches cut nothing other than natural subsoil and in profile were steep-sided, on average *c* 0.45 m wide at the top and *c* 0.25-0.3 m at the bottom, which was usually rounded. The bottoms were cut more or less horizontally into the slight eastward slope to an average depth of *c* 0.7 m. Where butt ends occurred they were generally rounded in plan, as were most corners. Into these trenches posts had been placed, usually centrally but occasionally to one side or the other; the recurrent centre-to-centre spacing in external walls appeared to be *c* 0.9 m (3 Roman feet?) while internal walls were often more closely studded, at as little as 0.4-0.5 m centres. All the foundation trenches were backfilled with excavated subsoil of varying kinds; where the type of backfill matched the contiguous subsoil, trench-edges were often difficult to trace, even where the wall-lines were clear from other evidence.

Within the backfill lay post-impressions, the shape of which became better defined with depth. In the outside walls these were generally square or rectangular, with average sizes of 0.10 m x 0.10 m and 0.12 m x 0.08 m respectively, while internal walls had apparently employed circular posts between 0.12 m and 0.14 m in diameter. Some post-impressions were misshaped, presumably by extraction of the posts; occasional larger post-impressions possibly marked the position of doorways, or implied a wider spacing of posts along a facade, such as that of Verandah 15 in the south Range. The post-impressions were filled with a variety of materials towards the top and a distinctive sandy loam lower down.

No outside walls from this stage remained above ground level, but evidence for the infilling of the internal frame came from the wall dividing Rooms 11 and 12 in the South Range, where a stub of clean orange-brown clay 0.2 m wide lay between and around the post-impressions, the bottom of the clay being level with the top of the backfilled foundation trench. Along the centre of the clay was a thin dark-brown band of loam, up to 0.02 m wide presumably the impression of wattle or lath infilling, sprung between the uprights as a key for the clay walling. Fragments of wall-plaster sealed against the foot of the wall by the Stage II concrete floors showed that the partition had been plastered on both faces in Stage I.

North Range

None of the foundation trenches of this range was seen above natural subsoil. The foundation trench defining the southern sides of Corridor 4 was constructed as one with the foundation trench of the eastern side of Alley 7 and can therefore be seen as part of the original layout.

The range comprised two large rooms (1 and 3) the eastern being particularly spacious; these lay either side of a smaller central room (2). They were all linked by a Corridor (4) on the southern side. The range measured 5.5 m north-south to the centres of foundation trenches; the east-west measurement was at least 23.7 m on the south, where the range ran beyond the eastern limit of excavation, but perhaps as much as 26.4 m on the less well defined, but possibly complete northern side. There, the northern side of Room C came to a rounded end west of a modern ditch and was not traced to its east, despite detailed spit excavation in search of anything that might have continued the alignment. From the east end of this foundation trench there was a gap of 5.8 m before the alignment was resumed by four postholes, alternately large (*c* 0.35 m diam) and small (*c* 0.15 m diam), but no more than 0.10 m deep. Any post-pits further east should have been seen in the extreme north-eastern corner of the excavation but none was found. Despite truncation and pitting of this part of the site, it is unlikely that foundation trenches like those of Stage I would have been entirely removed; nor are they likely to have been missed during excavation. It may therefore be suggested that the north-east part of the North Range was planned as an open-sided structure.

Rooms 1 and 2 were separated by a shallow foundation trench of rectangular profile, only 0.15 m deep, cut into clean subsoil. The post-impressions along it were irregularly spaced, though similar in size to others of the range. Despite these variations, contemporaneity with the rest of the Stage I foundation trenches was suggested by the similarity of their fills and the rounded shape of the junctions at the north and south; it seems unlikely that a later intrusion would have produced re-entrant corners of the kind shown in Fig 6.

Considerable difficulties were experienced in the excavation of the partition between Rooms 2 and 3. The fills of three irregular pits (F6, F7, F8) which formed an apparent division between the rooms, were barely distinguishable from the surrounding subsoil. The pits contained no finds apart from a worked flint, and the possibility of their being prehistoric cannot be discounted. They formed a north-south alignment which roughly complemented rounded ends on both of the corridor foundation trenches. The alignment was continued by a linear foundation trench (F10) which would have bisected a gap in the southern foundation trench of the range had it not stopped three-quarters of the way across Corridor 4. The gap might represent two doorways leading either side of the Corridor dividing-wall.

Few of the features in the North Range shown on Fig 6 could be allocated securely to Stage I and they are discussed below (p 12). A tiny remnant of plain white wall-plaster, however, appeared to be lying *in situ* inside the southern foundation trench of Corridor 4, suggesting the walls may have been plastered from an early stage.

On the northern side of Room C a post-pipe (FB) 0.17 m diameter and 0.27 m deep lay in a post-pit cut by the northern foundation trench for the North Range; it may belong to the time the range was being built. The plan of Stage I makes it clear that not all post-positions were detected and only those considered certain are marked on the plan.

Two other features cutting the natural subsoil resembled post-pit FB; one (FA) was not excavated but the other (FC) contained a post-pipe 0.14 m diameter and 0.25 m deep. Two unexcavated stakeholes lay to the east of FA. Running diagonally across Room C towards its north-west corner was a shallow gully 0.03 m deep and 0.15 m wide, filled with pink clay and sooty loam. Marginally overlapping this and respecting the position of the post-pipe in FC was a small patch of sandy loam. The features present no readily-

STAGE I

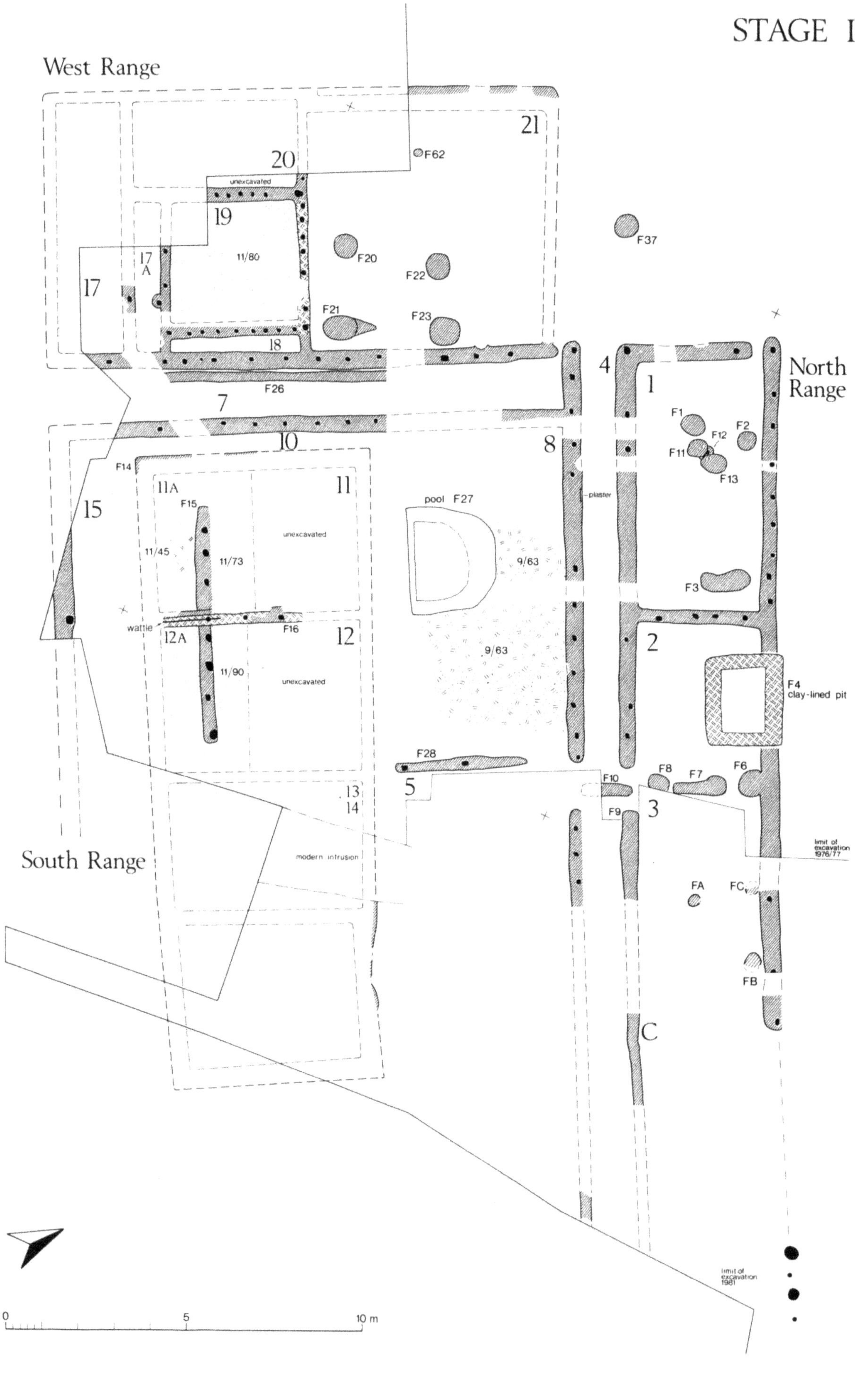
West Range
North Range
South Range
1
2
3
4
5
7
8
10
11
11A
12
12A
13
14
15
17
17 A
18
19
20
21
C
F1
F2
F3
F4 clay-lined pit
F6
F7
F8
F9
F10
F11
F12
F13
F14
F15
F16
F20
F21
F22
F23
F26
F28
F37
F62
FA
FB
FC
pool F27
unexcavated
modern intrusion
plaster
wattle
11/80
11/45
11/73
11/90
9/63
limit of excavation 1976/77
limit of excavation 1981
0
5
10 m

comprehensible pattern.

The South Range

This range comprised four large rooms, the western pair being initially fronted by vestibules (11A and 12A). A gap between Room 11 and Alley 7 must have been used as a Corridor (10). The foundation trench delimiting the southern side of the range contained a post-impression 0.2 m square, perhaps reflecting a wider spacing of more substantial posts along an open verandah, which presumably joined with Corridor 10. Only a small part of the eastern pair of rooms was excavated and time did not permit the careful examination of the subsoil which revealed the foundation trenches of the vestibule and verandah in the western pair; the absence of evidence for these rooms in the eastern pair is not, therefore, conclusive.

The general layout of the buildings suggests that the wide verandah along the southern side of the range may have represented part of the southern facade of the complex. Excluding the conjectured Corridor 6, the range measured 8.25 m x 16.5 m to the centres of its foundation trenches.

The surviving foundation trenches were seen cutting natural subsoil only, and were partially covered by the distinctive concrete floors of Stage II. The earliest remaining foundation trench (F15) separating Rooms 11 and 12 from their vestibules 11A and 12A, was cut by a north-south trench dividing Rooms 11 and 12 (F16) but should not necessarily be seen as a feature of a different phase; the earliest detected layer in Room 11 (11/73) ran over the fills of both F15 and F16, but in the case of the latter abutted the clay wall-stub, and in the former, did not cross the wall-line suggested by the post-impressions (between 0.10 m and 0.18 m across, perhaps as a result of disturbance on removal). Both wall-trenches must therefore have been in use simultaneously at some time during Stage I, although it is theoretically possible that F15 also existed alone at an even earlier stage, with a floor of natural subsoil.

The boundaries of Rooms 11 and 12 and their vestibules were implied by the limits of the floors associated with the internal partitions, the foundation trenches themselves having been totally removed by the Stage III rebuilding work apart from a sliver on the south-western external corner of Room 11 (F14). Two rooms occupied the space to the east of Room 12 in Stage III, one of which was subdivided (Rooms 13 and 14); these may represent rebuilding of a timber structure *in toto*, the only trace of the earlier eastern pair of rooms being a sliver of a linear feature, presumably a foundation trench on the northern side of Room B.

Preserved floor surfaces of Stage I within the South Range were few. The original floors in three of the rooms consisted of loam (11/73, 11/90), and a patch of clayey loam remained in the other (11/45). Before the alteration during Stage I (p 12), Room 11 was resurfaced with clay and Vestibule 11A with gravelly clay. No such resurfacings were traced in Rooms 12 and 12A, though there was a large patch of charcoal in the centre of Vestibule 12A. Grey discolourations on the surface of natural subsoil in Verandah 15, Corridor 10 and Room B, presumably from their use in Stages I and II, were noted but could not be examined in detail.

West Range

As with the South Range, the foundation trenches of the West Range were seen cutting natural subsoil only, and were sealed in part by concrete floors firmly attributed to Stage II. Their use in Stage I cannot reasonably be doubted.

The northern half of the range was apparently taken up by a single large room (21) to the south of which were two smaller rooms and a narrow space perhaps best interpreted as a cupboard (20, 19, 18); fragmentary evidence further south hints at an arrangement of vestibule and verandah (17A, 17) similar to that of the South Range. The southern boundary of the range lay beyond the limit of excavation, but if it matched that assumed for the facade of the South Range, the overall dimensions of the West Range can be placed at approximately 7 m x 14 m to the centres of the wall trenches.

The northern and southern foundation trenches of Rooms 19 and 20 were a typically narrow, the posts being more closely spaced than in external walls. The northern side of Room 19 retained a stub of clay walling, resembling that between Rooms 11 and 12 in the South Range.

The northern limit of the West Range was problematical. The western foundation trench was partially cut away by late walls, but enough of its northern end remained to show the curve of either a butt end or an eastward-turning corner. Opposite this, the eastern foundation trench, although partially cut away by a later feature, also displayed what appeared to be a butt end. As the pattern of trenches at the east end of the North Range readily shows, this need not imply the absence of a cross-wall, though no evidence was detected for a trench between these two points. If the range was not open-ended, as this implies, any wall-trench could have either been missed in the difficult subsoil, or more probably, completely removed by a foundation trench (F42) cut along the same line in Stage II.

Surviving Stage I floor surfaces in the West Range were observed only in Room 19, where a brown loam, partially discoloured grey (11/80) was sealed beneath the Stage II concrete floor, and in Room 20, where a similar brown loam beneath the Stage II concrete may

Figure 6: Plan of Stage I. scale 1:150

also have been a Stage I floor.

External Areas

Between the North, South and West Ranges were two external areas. Alley 7, about 1.5 m wide between its flanking wall-faces, lay between the South and West Ranges and continued across the western end of Yard 8, which occupied the space between the North and South Ranges. The eastern wall of Alley 7 provided a clear boundary to Yard 8 on the west but the search for an eastern boundary presented problems similar to those encountered in excavating the partition between Rooms 2 and 3. None was observed, however, within the excavated area. Minor colour-variations in the subsoil suggested the presence of a linear feature 0.35 m-0.40 m wide (F28). On excavation the edges and bottom of F28 were difficult to establish with confidence, but the post-impressions indicated on Fig 6 were definite and provided evidence for a division of Yard 8 in Stage I; the western part measuring 5.25 m x 9 m and the eastern part 5.25 m by at least 8.25 m. The only surviving feature in Alley 7 was a shallow U-sectioned gully (F26), parallel with and close to the eastern wall of the West Range, interpreted as a storm-water gully. The stratification in Yard 8 is discussed separately (p 16) because of uncertainty in its phasing.

Alterations during Stage I

The only alteration to the complex definitely attributable to Stage I was the removal, in the South Range, of the partition dividing Rooms 11 and 12 from their vestibules. This is shown to have occurred before Stage II by the sequence in Rooms 11 and 12, where additional floor surfaces of pebbles, gravel and grit beneath the Stage II concrete floor ran uninterrupted across foundation trench F15, showing that the partition had been removed before the end of Stage I.

Stratification possibly of Stage I

The following paragraphs describe layers and features of the North and West Ranges and External Areas (Fig 6) which were sealed by stratification of Stage III or later, but which bore an indeterminate strategraphic relationship to features known to be of Stage I. There was no stratification of this category in the South Range.

North Range

Four pits in Room 1 were filled with redeposited natural subsoil, without finds. Three (F1, F2, F3) were seen cutting natural subsoil only, but the fourth also cut a post-pit (F12); pits F1-3 seem likely either to have been prehistoric or to have belonged, with post-pit F12 and pit F11, to a time early in Stage I before occupation refuse had accumulated about the site. A fifth pit (F13), also cutting post-pit F12, contained a few small body-sherds of undistinctive Roman pottery.

Interrupting and aligned with the northern wall of Room 2 was a tank set in a clay-lined pit (F4) cut into the natural subsoil. In its north-west corner was the everted lip of a lead pipe, presumably an inlet, which protruded slightly from the clay as if to accommodate a timber lining. The tank's fill of homogeneous rubbly loam was cut by ditch F97, itself out of use by Stage IV when it was cut by ditch F300. The placement of the tank later in the sequence than Stage I is implausible because its construction would then have involved the undercutting of the northern wall of Room 2; if constructed at a later date the pit would surely have been placed entirely within the room, so as to leave the external wall undisturbed. In Stage I any northern wall as this point could presumably have been based on a sole-plate resting on the northern side of the pit's clay lining.

Less securely stratified features in Room C are described below (p 20).

West Range

Within Room 21 and sealed by the debris of demolition work between Stages II and III were four pits (F20-23), seen cutting natural subsoil only. In the absence of linking stratification connecting them to walls their phasing is uncertain, although their uncontaminated fills suggested they belonged to an early time in the life of the complex. Sediment analysis of a sample from pit F22 offered no real guide to its depositional origin.

External Areas

On the southern side of Yard 8 was a semi-circular pool (F27), cut into the natural subsoil and shown by its infilling to have been out of use by the start of Stage IV (at the latest). Stratigraphically, therefore, it could have belonged to any part of Stages I-III but its semi-circular shape seems appropriate to its interpretation as a formal element in the carefully designed layout of Stage I.

North of the West Range

To the north of the West Range was a solitary pit (F37) sealed by the clay beneath the Stage II concrete floor of Room 23. Its uncontaminated fill suggests that if not of a pre-complex period, it probably belonged to a time early in Stage I.

Dating evidence for Stage I (PVW)

The samian evidence suggests occupation near the excavated site from *c* AD 90, and it is supported both by the large number of late first/early second century coarse pottery vessels present on the site, and by the date of the material from earlier excavations at Pentre Ffwrndan (p120). The construction of the complex cannot, however, date from as early as AD 90 because amongst the small amount of material from Stage I were sherds of Black-burnished ware (nos 26-7) and Central Gaulish samian (not illustrated) of form 35. Black-burnished ware is unlikely to have been imported into the area before *c* AD 120. Therefore, if nos 26-7 do belong to the initial construction of the complex it must have been built after *c* AD 120. An early Hadrianic date for Stage I may therefore be suggested.

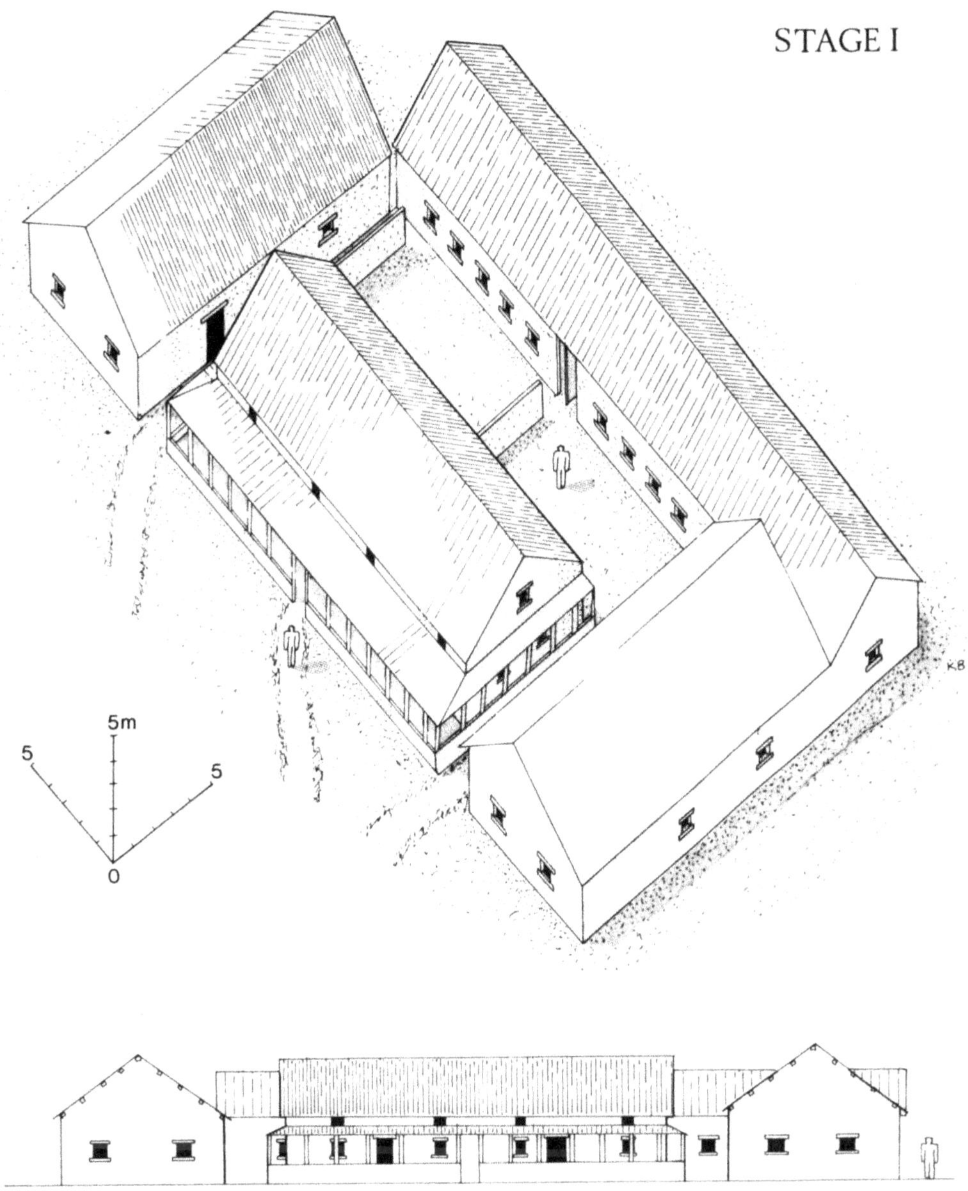

Figure 7: Tentative reconstruction of Stage I

STAGE II

The changes which define Stage II (Fig 8 and pl I) involved the building of two extensions, each consisting of two rooms; one lay to the west of the West Range (West Extension; Rooms 24, 25) and the other in the angle between the North and West Ranges (North-west Extension; Rooms 22, 23). The foundation trenches cut those of Stage I and differed fractionally in orientation, as well as in fill and post-spacings (Room 19 in Stage I excepted). The extensions were only sealed by Stage IV stratification, but an argument is presented (p 14) for their independent phasing in Stage II. In addition, a small room (26) was inserted in the south-eastern corner of the single large room (21) which occupied the northern half of the West Range in Stage I. There is evidence that this room in turn was demolished before the changes attributed to Stage III, and also arguments to suggest that Room 21 of Stage I had become an open yard in Stage II (Yard 21A).

The two extensions were almost certainly built as separate structural elements, although their foundation trenches were shown by their continuity in plan and filling to belong to a single phase of construction.

They varied in several respects from the foundation trenches used in Stage I. The average depth measured from the surface of the subsoil was *c* 0.15 m less, at 0.55 m. The width at the top varied between 0.35 m and 0.45 m. The main difference, however, lay in the size, shape and spacing of the post-impressions, which were round, smaller than those in Stage I (*c* 0.10 m diameter) and more irregularly but closely spaced (0.3 m centre-to-centre on average). The apparently unsquared posts of Stage II suggest that, in contrast to Stage I, timbers arrived at the site unprepared. Strangely, the pattern of post-impressions in the new foundation trenches makes no allowance for doorways. Remnants of clay walling around Room 23 resembled the infilling between the uprights of Stage I, although there was in this case no remaining evidence for the use of wattles or laths; it may be, however, that the close spacing of posts made wattling unnecessary.

West and North-west Extensions

The eastern boundary of the North-west Extension was formed by the North Range, and the southern boundary by a new foundation trench along what is taken to have been the original north end of the West Range. The North-west Extension measured 6.75 m x 4.25 m to the centres of the wall-posts. The West Extension was bounded on the east by a redefined wall-line dividing it from the West Range; from the centres of its wall-lines it measured 8.25 m east-west by a minimum of 3.15 m north-south, Fig 8 showing a purely notional continuation beyond the southern limit of excavation.

The new foundation trenches, constructed as one, cut earlier foundation trenches in three places; the two butt-ended foundation trenches of the North-west Extension cut away the greater part of the fôundation trench which had formed the boundary of the West Range in Stage I. The fills of the new foundation trenches differed from those of Stage I in that they were contaminated with pink clay, plaster flecks, pottery and bone, while the backfills of the earlier foundation trenches were uncontaminated.

Room 23 was floored with a distinctive white pebbly concrete (10/80); tiny fragments of crushed pottery or tile in the surface, thrown down when the laid floor was still wet, may have been intended to add colour. The concrete sealed a thick make-up of clay, without discolouration or detectable wear on its surface. Both the concrete and the clay sealed the edges of the foundation trenches around the room and ran up to the surviving clay wall stubs. Elsewhere in the complex (Rooms 11, 12 and 19) identical concrete floors directly sealed Stage I stratification; the absence of any detectable floor-finish beneath the concrete in Room 23 suggests that it was contemporary with the Stage II construction work.

In Room 24 patchy remnants of a gritty mortar floor (10/97) overlay the edges of the northern foundation trench. In the very north-west corner, at the junction of wall and floor, lay a fragment of quarter-round moulding (10/96). In the same corner of the room was a shallow pit, vertical-sided and flat-bottomed (F44); its obviously later fill of midden material (like that of F43, p 16) and its precise positioning in the corner of the room suggest that it may have held a tank connected with the use of the room, rather than with its construction.

Fragments of painted plaster from the demolition debris of both extensions, and from pits cutting the debris, suggested that Rooms 23 and 24 at least, had been given a decorative finish (pp 88-90).

South and West Ranges

Concrete floors, identical to that in Room 23, were found in Rooms 11 and 12 of the South Range and in Rooms 19 and 20 of the West Range. They are taken to mark contemporary points in otherwise separate sequences of stratification. In Rooms 11, 12 and 19, the concrete was cut by foundation trenches of rebuilding work (Stage III) which was clearly earlier than the erection of the first stone range (Stage IV). On this basis the concrete floors and their related stratification are placed independently as Stage II.

The poor state of preservation of the concrete floor in the southern part of Room 11, above the former Vestibule 11A, is thought to have resulted from leaching of the cement element due to the varying permeability of the underlying Stage I floors. The concrete in Room 11 (11/9) was a little coarser than that of Room 12 (11/2) and also a little thinner, though the surfaces lay at the same absolute level. In the north-west corners of Rooms 11 and 12 were patches of charcoal, devoid of finds. This could have accumulated at any time between Stage II and the abandonment of the floors at the end of Stage IV, but, interpreted as waste from braziers, they mark a function appropriate to the rooms in all stages.

In Room 19 the concrete (11/20) was almost intact and founded on a make-up of yellow clay. Investigation of Room 20 was restricted to a small *sondage c* 1 m square at the bottom of which was an identical concrete floor (3/100) resting on brown rather than yellow clay.

West Range

Room 26 was inserted in the south-eastern corner of former Room 21, using pre-existing walls for its southern and eastern sides and new foundation trenches on the north and west; it measured 4.5 m x 3.75 m to the centres of the wall-lines.

At the north-east corner of Room 26 its new

Figure 8: Plan of Stage II, scale 1:150

STAGE II

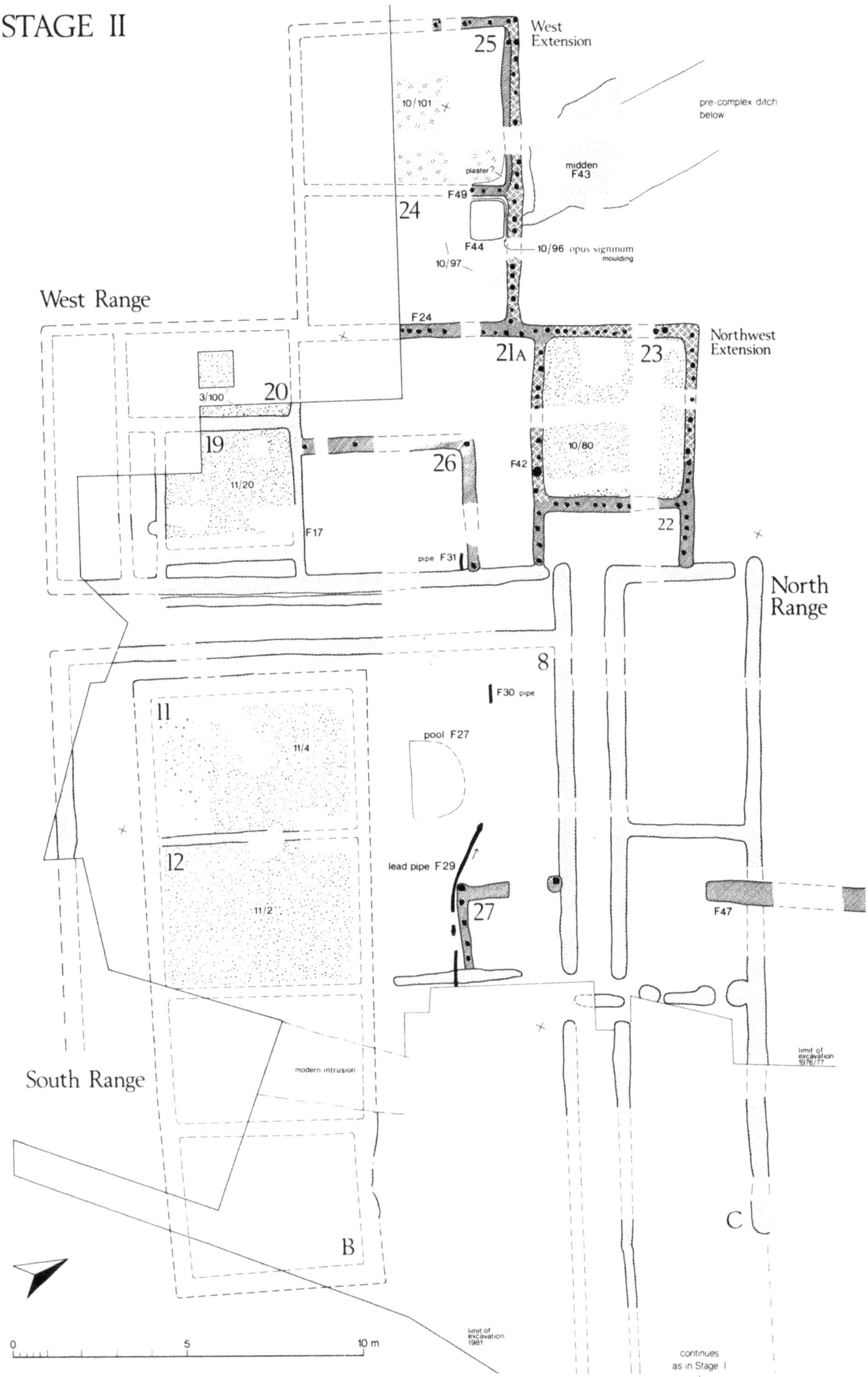
West Extension
25
pre-complex ditch below
10/101
plaster?
midden F43
F49
24
F44
10/96 opus signinum moulding
10/97
West Range
F24
Northwest Extension
21A
23
3/100
20
19
26
10/80
F42
11/20
F17
22
pipe F31
North Range
8
F30 pipe
11
11/4
pool F27
12
lead pipe F29
11/2
27
F47
limit of excavation 1976/77
South Range
modern intrusion
C
B
limit of excavation 1981
0
5
10 m
continues as in Stage I

foundation trench clearly cut that of the Stage I building: at the south-west corner there was no clear distinction between the trench fillings, but the new wall-trench was 0.05 m shallower, with a southernmost post-impression which duplicated one of the Stage I timbers. Unlike Stage I foundation trenches, those of the newly constructed walls contained lumps of pinkish clay; in part they were sealed by debris from the demolition work that took place between Stages II and III.

Consideration of the superstructure of the Stage II extensions and of Room 26 firmly suggests that the now L-shaped remainder of Room 21 should more correctly be seen as unroofed; henceforth it will be referred to as Yard 21A. Features in this yard possibly belonging to Stage II are described in the next section.

External Areas

In the absence of preserved stratification in Alley 7, it is assumed that gully F26 may have remained in use in Stage II. Pool F27 in Yard 8 almost certainly remained in use, until its eventual backfilling early in Stage IV.

Stratification possibly of Stage II

West Range

To the north of Room 24 was a widespread deposit of sandy loam (F43) (Fig 6). Despite its apparent homogeneity two concentrations of bone and shell formed tip lines, and this, together with a high content of kitchen refuse, suggested it was a midden. Its survival at this point may have been enhanced by its position above subsidence into the pre-complex ditch (F0) (Fig 18). The presence of thin patches of similar occupation refuse above, but noticeably not beneath, the floor surfaces of the adjacent rooms, suggests that the midden only began to accumulate after the North-west Extension had been built. Its usage may have continued throughout Stages II and III, both the midden itself and the rooms to the south and east only being sealed by debris from the demolition which occurred between Stages III and IV.

External Areas

The early stratification in Yard 8 was not examined in detail, leaving uncertainties about relationships between the main features. Lying to the north and north-east of pool F27 was a layer of orange-brown clay-loam, up to 0.2 m deep and apparently formed of redeposited natural subsoil. Within the same area of the yard was a lead pipe (F29) comprising three pieces connected at poured lead joints, one of which overlay a pit (F87), tentatively ascribed to Stage I; the relationship of pit and pipe to the clay-loam was not observed. The clay-loam, on the other hand, clearly sealed the foundation trenches of a small room (27) built into the north-east corner of Yard 8, the foundation trenches themselves being contaminated with flecks of white plaster and lead oxide, suggesting the room did not belong to a primary stage of construction. Two of the post-impressions in its west wall, however, were clearly visible in the surface of the clay loam, which would therefore appear to have been laid after the timbers of Room 27 had already been placed in position, though probably during the same general operation. When found, the western end of the lead pipe stood exposed above the surface of natural subsoil, an unlikely situation while it was still in use; the rigours of a northern climate surely imply that it would have had the protection of some encasing layer such as the clay-loam noted above. The pipe lay close alongside the south wall of Room 27, but changed direction outside the south-west corner as if to avoid both the room and Pool F27, on a presumed course towards smaller remnants of pipe further west in Yard 8 and in the north-east corner of Room 26 (F30, F31).

It is argued above (p 14) that Room 26 came into existence at the beginning of Stage II. It is clear, however, that it had gone out of use by Stage III, when the area it formerly occupied was converted into a single open space with Yard 8. It seems likely from the form of the gully in which the pipe lay at its observed western end (F31) that it rose above ground level at this point, presumably to supply water to the north-east corner of Room 26. On the basis of its apparent connection with Room 26 (out of use by Stage III) the pipe seems unlikely to be a Stage III addition, and it is assigned with reasonable confidence to Stage II, along with the clay-loam deposit and the walls of Room 27.

The Room 27 wall-alignments did not exactly match those of the other rooms. The trenches were shallower (0.35 m in clean subsoil) than most of Stages I-III and were square in profile. In the southern trench were six rectangular post-impressions measuring 0.10 m x 0.15 m, at between 0.35 m and 0.60 m centres. Apart from the south-west corner post, only one definite post-impression was traced in the western foundation trench, the northern end of which became uncertain where it coincided with pit F62 (Stage I). A further post-impression, however, was found in a post-pit hard against and clearly cutting the Stage I wall-trench on the North Range Corridor; this may be taken as a continuation of the west wall of Room 27, perhaps even denoting a door-post.

Dating evidence for Stage II (PVW)

The latest pieces from this stage (nos 32, 35-6) are Hadrianic-Antonine but none need necessarily be later than Hadrianic. The assemblage is comparatively small but seems similar to that found on military sites thought to have been abandoned at the time of the Antonine advance into Scotland.

Demolition between Stages II and III

Prior to the rebuilding described as Stage III substantial parts of the South and West Ranges were demolished. Comparison of Figs 8 and 9 shows clearly the removal of Room 26, the cutting back of the east wall of the West Range, and the amalgamation of Yard 8 with Alley 7 and Yard 21A to form a single space. At the same time the verandah and external corridors of the South Range were removed, leaving only the central rooms, which were themselves rebuilt on the same plan.

Material evidence for demolition between Stages II and III was scant, being confined to Room 26 only. The removal of other rooms has been inferred from the planned area of a Stage III clay deposit (11/27, Fig 9).

South and West Ranges

Within the northern part of Room 26 was a mixture of orange clays with loam and much white wall-plaster (10/92A, Fig 9), interpreted as demolished walling. The debris was sealed by two patches of charcoal, in turn sealed by a primary floor of Stage IV. The south-western corner of the room, including the original eastern end of foundation trench F17, was also sealed by the clay (11/27).

Clay deposit 11/27 completely sealed the foundation trenches forming the eastern side of Corridor 10, the southern side of Verandah 15, the northern, southern and eastern side of Cupboard 18, and part of the eastern wall of Room 26. All walls with foundation trenches completely sealed beneath the clay must have been demolished before its deposition. In contrast, there was no direct evidence for the demolition of the northern end of Alley 7 prior to Stage III, as opposed to Stage IV; but the effective widening of Alley 7 between the South and West Ranges in Stage III, suggests that the layout of the original alley had already been abandoned.

On the west of Room 11 and the south of Rooms 11 and 12, the outer sides of foundation trenches, which, on the inside had cut distinctive Stage II concrete floors, were also sealed by clay deposit 11/27. Thus, it is clear that these rooms were demolished before being rebuilt on the same plan in Stage III. Rooms B and 13/14 were also rebuilt in Stage III and it is assumed that, in preparation, the earlier pair of rooms was demolished at the same time as the rest of the range.

STAGE III

After the demolition work described in the previous section, the central rooms of the South Range and parts of the West Range were rebuilt (Fig 9). No alterations attributable to this Stage were detected in the North Range or West and North-west Extensions.

The method of construction varied slightly from those of Stages I and II. The foundation trenches were uniformly wider, measuring 0.7 m across at the top and 0.35 m at the bottom. Instead of being filled with clean or nearly clean subsoil, they were now heavily contaminated with plaster and clay, and contained a variety of finds. Post-impressions were clearly defined, both in the fills of the foundation trenches and in the remnants of clay walling above them. They were sub-circular and often larger than those of Stages I and II, measuring up to 0.20 m across. The centres were generally 0.3 m to 0.5 m apart, but some wider spacings may mark the position of doorways. The impressions penetrated the full depth (0.7 m) of the foundation trench, demonstrating the continued absence of sole-plates. Around the South Range, lying above and level with the top of the foundation trench fills were remnants of clay used as infilling between the timber uprights. The height varied from 0.03 m to 0.20 m; where not disturbed by the imposition of later deposits the width was about 0.20 m, as in Stages I and II. Along the centre of the clay stubs in the northern wall of Room 12, the western wall of Room 11 and the wall dividing Rooms 13 and 14 were thin bands of loose brown loam between 0.01 m and 0.02 m wide, presumably marking the positions of horizontal wattles interwoven irregularly between the uprights.

The bases of the external walls of the South Range were sealed by a fillet of greenish clay up to 0.2 m wide, forming an angle of about 45 degrees between the base of the wall and the surrounding ground surface. The fillet along the north wall of Room 12 incorporated a line of amphora sherds laid immediately against the outer face of the surviving wall-stub, while that along most of the southern and western walls of the range overlay a line of pebbles placed in a similar position. The incorporation of the pottery and pebbles makes it likely that the fillet was purposely laid in this manner, and did not simply comprise the remnant of an external weatherproof rendering which had been washed to the foot of the wall by the elements; no greenish clay of the type used in the fillet was found in the demolition debris of this range.

Plaster attached to the inner faces of the wall-stubs in the South Range indicated how the new walls had been finished in Stage III. Much of the painted plaster found in the debris resulting from the later demolition of these rooms (between Stages VI and VII) must have derived from Stage III or later.

South and West Ranges

Foundation trenches around Room B and 13/14 were continuous with those of Rooms 11 and 12, which butt the edges of the distinctive Stage II concrete floors.

The outer edges of the new wall-trenches were overlapped by clay make-up (11/27), which butted against wall-stubs or, where these did not remain, respected the alignment of the walls suggested by the post-positions. The clay make-up was sealed by the fillet of external weather-proofing where that was present and in the north was cut by the foundation trench for the southern wall of the Stage IV North-west Range.

Apart from the wall dividing Rooms 11 and 12, the South Range was completely rebuilt as four, equally large rooms, the more westerly of the eastern pair being subdivided into two equal parts (Rooms 13 and 14) by an east-west wall which survived either side of a modern drain. The rounded north-eastern corner of Room B marked the eastern end of the range.

As mentioned above, the Stage III concrete floors were cut back during digging of the Stage III foundation trenches, the gap between the ragged edges and the external walls then being patched with what appeared at the time of excavation to be gravel, but which may have been laid originally as concrete, the cement fraction having subsequently leached out.

Too little of Room 14 was investigated to distinguish any flooring. In Room 13 an orange clay lay beneath a pebbly loam (11/38). Even though these were only sealed by debris from the demolition of the South Range between Stages VI and VII, they have been described in Stage III because they were the first deposits to overlap the edges of the surrounding foundation trenches.

Similar arguments apply to the later of two features in Room B. The earlier feature, partially obscured in plan by detritus from the later feature, was half-sectioned longitudinally (north-south). It was half figure-of-eight, the southern end being slightly deeper (0.20 m) than the northern (0.16 m). Its fill was a matrix of finely broken charcoal and brown clay-loam with a concentration of charcoal above the depression at the southern end.

The unexcavated south-eastern corner was cut by a circular feature, 0.25 m deep, extended to the north by a smaller, shallower, sub-rectangular cut. The base of the feature was covered with a very thin layer of charcoal or soot (2 mm thick), though the sides of the pit itself showed no hardening or burning. The soot was sealed by a dense pink clay in which were three flat-laid pieces of Hawarden Sandstone, themselves covered by a bowl-shaped layer of pink clay laid only around the outer edges, thus leaving a piece of sandstone *c* 0.10 m diameter, still exposed. In this bowl-shaped depression were two layers of lead (0.02 m thick) separated by a lens (0.01 m thick) of clay and lead oxide. So closely did the lead fit the bowl-shaped contours that it had obviously been deposited in a semi-molten state. The slightly concave upper face of the lead was covered by a mixed layer of charcoal and clay, 0.05 m thick, which did not, however, fill the feature to contemporary ground level. This was done around the outer edges by a dense clay with mottled inclusions, which like the clay below left empty a bowl-shaped central space. This and the feature as a whole was levelled off by sandy clay and charcoal, flecked with lead oxide.

Whilst the function of the earlier feature cannot be securely identified, the later one was clearly a bowl furnace. The floor of the room around it was covered with charcoal and soot, flecked with lead-oxide, all mixed with pink and red clays. This sealed the edges of the foundation trenches on the northern and southern sides of Room B.

In the West Range, all the walls around Cupboard 18, and the east wall of Room 17, had been demolished before Stage III. Most of Room 19 and presumably Room 20 remained standing, however, to be repaired by a new Eastern wall, which was shown by the plan of a related floor-finish (11/82), to have turned west to form the new southern end of Room 17. On the basis of later boundaries in the same position, this may also have been the southern limit of the range. The removal of Room 26 converted the northern wall of Rooms 19 and 20 into the external wall of a much-reduced West Range, the area to the north now becoming a part of the enlarged open space formed by the amalgamation of Yard 21A, Alley 7 and Yard 8.

As in Rooms 11 and 12 of the South Range, the concrete floor of Room 19 appears to have remained in use in Stage III, the gap between its cut-back edge and the new eastern wall being made good with a gravelly deposit (11/49) similar to that in the equivalent position in the South Range (11/81). Room 17 was finished with firm gravelly clay-loam (11/82) which remained in patches, some overlapping the edges of the new southern and eastern foundation trenches.

North-western extension

In Room 24, the possible tank ascribed to Stage II (F44) was eventually filled, probably in Stage III, with material resembling that of the midden immediately outside the room to the north (F43).

External Areas

In Yard 8B, furnace F69, overlay the area formerly occupied by Room 27 and was itself covered by debris from demolition work pre-dating Stage IV. The feature incorporated pieces of Hawarden Sandstone of the kind commonly used in Stage IV. In its upper parts was a little lead oxide and a large lump of lead; lead oxide was used elsewhere in the complex, however, as flooring and filling material and its presence here need not indicate the use of the furnace for lead-working. The furnace was covered by a charcoal deposit, 0.03 m to 0.10 m deep, containing molten lead lumps, greenish glass-like globules and other finds.

Above the former Room 26, in the much enlarged central space now designated Yard 8A, the demolition debris of the room (10/92A etc) merged with clay 11/27, suggesting that the debris may here have been used as a surface in its own right.

In the widened Alley 7 (now designated Yard 7A) the clay deposit had subsided slightly into the assumed storm water gully F26 (Stages I and II). The resulting

Figure 9: Plan of Stage III, scale 1:150

STAGE III

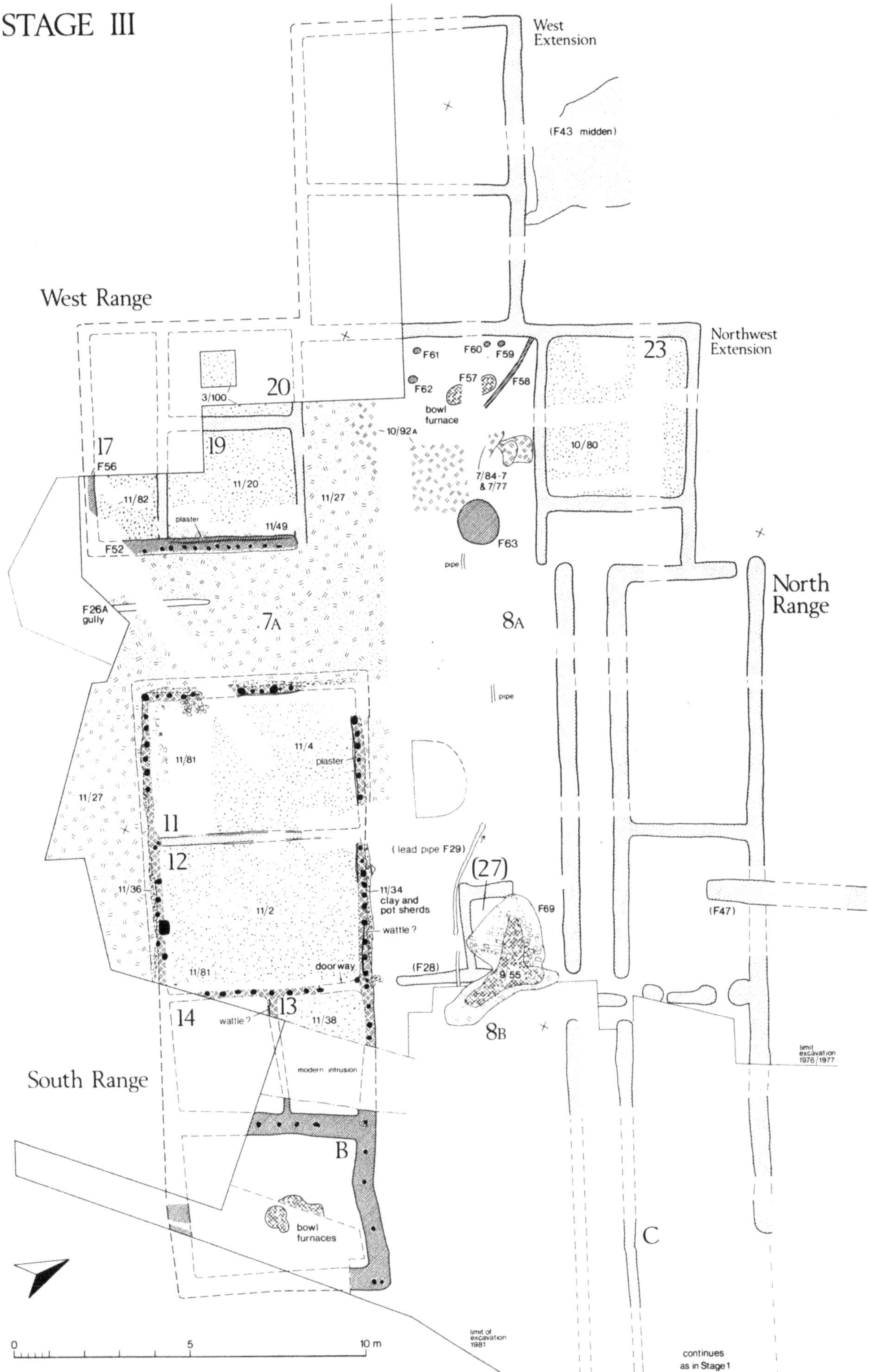
West Extension
(F43 midden)
West Range
Northwest Extension
23
20
3/100
F61
F60
F59
F62
F57
F58
bowl furnace
10/92A
17
19
F56
11/82
11/20
11/27
10/80
7/84-7 & 7/77
plaster
11/49
F52
F63
pipe
North Range
F26A gully
7A
8A
pipe
11/81
11/4
plaster
11/27
11
(lead pipe F29)
12
(27)
11/36
11/34 clay and pot sherds
F69
11/2
wattle ?
(F47)
11/81
doorway
(F28)
9/55
14
13
wattle ?
11/38
8B
limit excavation 1976/1977
South Range
modern intrusion
B
bowl furnaces
C
limit of excavation 1981
0
5
10 m
continues as in Stage 1

depression was filled by charcoal and sealed at its northern end by two patches of silty loam and charcoal; these were cut by three shallow scoops filled with a continuous charcoal deposit. In Yard 7A was gully F26A.

Dating evidence for Stage III (PVW)

There is only a small amount of pottery from this stage, but it may be considered together with that from the Stage II-III midden (nos 37-47) and the sherds from the demolition between Stages III and IV (nos 58-69). The latest pieces are from a constructional context (nos 51-3) and date to the mid to late second century and from demolition (no 64). These suggest that this stage may be of Antonine date.

Stratification of Stages I, II or III

Fig 9 includes several features which cut nothing other than natural subsoil but were sealed by the debris of demolition work undertaken prior to Stage IV. All must therefore belong to Stages I-III, though their precise allocation is a matter of uncertainty.

The most distinctive feature (F57) appeared to have a figure-of-eight shape, although it was bisected by a Stage IV foundation trench. Its shape, burnt sides and charcoally fill suggested that it may have been a bowl furnace, a speculation supported by sediment analysis which suggested iron-working. Above the backfilled furnace were three layers of charcoally loam.

To the north was a gully filled with loose loam (F58); it was not detected to the south of a point where it was cut by a Stage IV foundation trench, and its origin may have lain adjacent to the furnace. The gully ran towards the north-west and presumably cut through the re-entrant angle between the West and North-west Extensions. It was not observed in the area beyond, where its continuation may have lain in levels removed without examination during the final stages of the excavation.

Two small pits (F59, F60), sharing a similar filling to that of the furnace may have gone out of use at the same time. South-west of the furnace, but not necessarily related to it, were two small oval pits (F61, F62) of unclear function, filled with charcoal-flecked redeposited subsoil.

In the eastern part of Yard 8A was a sub-rectangular pit (F63), 0.10 m deep, filled with charcoal and small burnt bones. Its western side was cut by the surviving part of another shallow pit filled with charcoal-flecked grey loam. In general character these fills resemble others referred to in this section, but there is no proof that they belong to the first three stages. They were sealed only by modern farm buildings that had truncated the eastern parts of the excavated area.

Demolition between Stages III and IV

Before the construction of the North-west Range at the start of Stage IV, the North Range, the north end of the West Range, and the West and North-west Extension were demolished following a fire (Fig 10). There was no evidence for change elsewhere in the complex apart from the inferred removal of the Room 19/20 partition wall in the South Range.

In all of the areas of demolition, dense charcoal deposits devoid of occupation refuse overlay Stage III surfaces or their related foundation trenches, while clearly respecting the wall stubs or wall alignments suggested by the post-positions. Above the charcoal and covering the former wall alignments were layers of orange clay 0.1 m to 0.15 m thick, sometimes reddened and mixed with much wall plaster. The clay was cut by the foundation trenches for the Stage IV North-west Range.

The charcoal beneath the clay and plaster was powdery and uncompacted, containing neither tile fragments nor nails; only one burnt timber remained in anything like its original form. Taken together the clay and charcoal are confidently interpreted as debris derived from the burning and subsequent demolition of the Stage III buildings. The jumbled character of the clay and plaster suggests that the walls were broken down piecemeal before being levelled, in contrast to later episodes of demolition in which walling appeared to have been pushed over in complete sections.

West and North-west Extensions

Charcoal deposits varied in depth between 0.01 m and 0.09 m, the most common depth being 0.02 m. One charcoal deposit was, however, 0.10 m deep and was immediately overlain by a group of amphora sherds, in some cases blackened on their upper surfaces, but in all cases burnt grey on the bottom. At one point on the east wall of Room 24 a narrow strip of charcoal spanned the full width of the foundation trench, presumably marking the position of a doorway.

North Range

Charcoal deposits in the North Range were patchier than those in the West and North-west Extensions; clay debris was recorded only in a small area at the south-west, and in Yard 8B.

External Areas

In Yard 8B the northern edge of the charcoal from the Stage III furnace (F69) was overlapped by a layer of clay containing plaster lumps, *c* 0.10 m deep, assumed to be the debris from parts of the demolished North Range. The northern edge of the clay was cut by the Stage V wall (F92).

Figure 10: Plan of demolition between Stages III and IV, scale 1:150

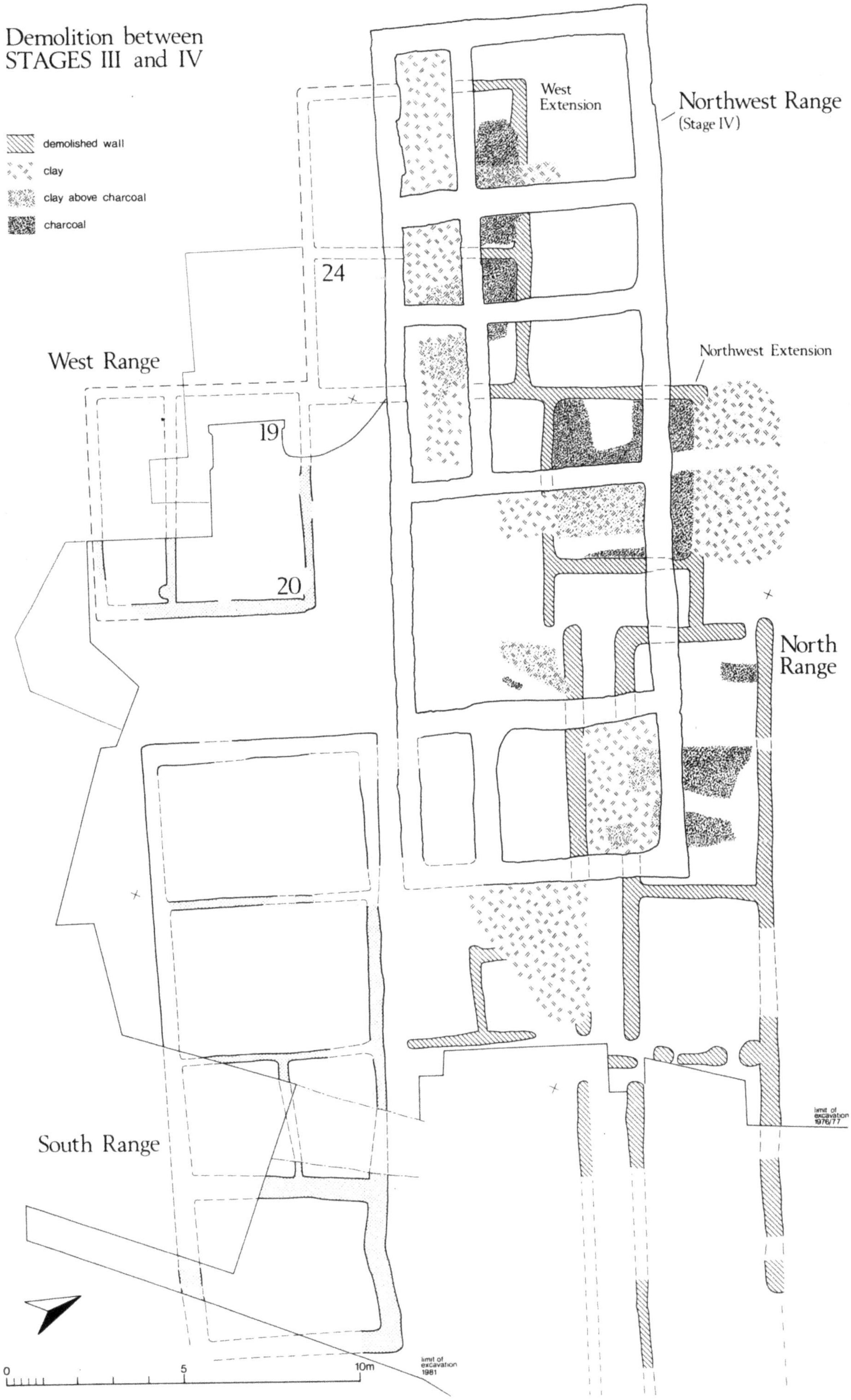
Demolition between
STAGES III and IV
demolished wall
clay
clay above charcoal
charcoal
West
Extension
Northwest Range
(Stage IV)
24
West Range
Northwest Extension
19
20
North
Range
limit of
excavation
1976/77
South Range
0
5
10m
limit of
excavation
1981

STAGE IV

The demolished Stage III rooms were replaced in Stage IV (Fig 11) by a stone-built North-west Range of five rooms, linked by a corridor on the southern side. The external dimensions of the range were *c* 8 m x 23.75 m. Most walls had been robbed to their foundation and many wall alignments were represented only by robber trenches. Detailed stratigraphic sequences were recovered from full excavation of Rooms 29-31 and Corridor 33. The deposits in Rooms 28 and 32 were rapidly half-sectioned and were then cleared at speed along with all deposits in Room 34, down to the remnants of the timber stages beneath. The stratified deposits within the individual rooms were isolated from one another either by robber trenches or by remnants of walls. Often, consisting as they did of nondescript patches of clay loam and occupation refuse, they were impossible to relate with confidence to individual stages in the development of the complex, apart from the first flooring materials.

The debris of the demolished Stage III rooms had a reasonably flat surface and may have been purposely levelled to form a base for the floors of the new North-west Range. The foundation trenches were cut to a depth of between 0.3 m and 0.4 m from the surface of the demolition debris, through the earlier deposits and into the natural sub-stratum. Where foundation trenches remained intact, they varied in width between 0.7 m and 0.8 m. Walls within them were founded on a single course of dressed stones, pressed firmly against the sides of the foundation trench; the stone used was a yellow-buff Carboniferous Limestone identified as originating in the region of Hawarden about 12.8 km to the south-west. The footing comprised stones laid in most cases as opposed facings of mixed headers and stretchers, the small central gap being made good with a mixture of pebbly loam and rubble. Above this, similarly laid Hawarden Sandstone had been used to form walls 0.6 m wide, preserved to a maximum height of five courses (0.40 m) along the southern wall of the range. No mortar remained between the courses, nor any gravel or sand which could have derived from decayed mortar (a fine brown loam between the stones in places could have been a post-depositional soil-formation). It seems probable, therefore, that the original walls were dry-built. The narrow spaces between the wall-faces and the edges of the foundation trenches had been backfilled with a mixture of yellow and orange clays with sandy inclusions, plaster and charcoal flecks, presumably derived from the Stage III demolition debris.

A post-socket (F87), in the southern wall of Room 31, implied the presence here of a doorway. There was no evidence that any of the new walls had been plastered; certainly they had not been so along the south side of Corridor 33, where both the Stage IV and later floor surfaces butted directly against the stonework, here alone in the North-west Range surviving above the contemporary ground level. Elsewhere, deeper robbing of the wall stones had removed any chance of recovering information on possible wall-finishes, either from the survival of plaster itself or from the recording of gaps left by its decomposition between floor finishes and matching wall-lines (as postulated in the case of the Stage V North Range (p 28).

North-west Range

The boundaries of all rooms were clearly marked by surviving walls or by differences between floor deposits on either side of robber trenches. The southern wall stood 0.75 m clear of the timber South Range and *c* 2.1 m clear of the much-reduced West Range.

Foundation trenches for the new Range cut the levelled demolition debris of the earlier timber West Range. In the northern part of Yard 7A and in the gap between the West and North-west Ranges two layers overlapped the construction trench for the southern wall of the North-west Range and abutted the wall itself; the second layer also directly abutted the stub of the west wall of the South Range. Above these lay demolition debris of the West Range, dating to before Stage V.

Within the range Room 28 was not excavated to this level apart from its south-east corner, where several burnt pieces of Hawarden Sandstone, lying flat in fine sandy charcoal and immediately above the demolition debris of the earlier timber building, probably represented a hearth (F66).

In Room 29 a thin layer of brown-black loam and one of clay-loam had been used to level off the subsidence of earlier deposits into the line of the pre-complex ditch, before the laying of a loam floor (6/58), which sealed the edge of the wall's foundation trench in the north-west corner of the room. The assumed Stage IV floor surface in Room 30 was patchy. Near the south-eastern corner, the eastern edge of a sandy clay-loam (6/42) overlapped the western edge of a sandy clay (6/49) which was in turn almost totally covered by two further layers: a group of stones (F68) perhaps forming some kind of a hearth base similar to that in Room 28, and a mortar-loam patch in a hollow alongside it (6/27). The northern part of the room was cut away by pits F67 and F139, ascribed to Stage V.

In Room 31, a pit cut the demolition debris of the earlier timber rooms and was sealed by the primary floor (10/66). This, the most substantial floor of the range, comprised an extensive layer of mixed clay and loam, more plastic than the clay of the demolition

Figure 11: Plan of Stage IV, scale 1:150

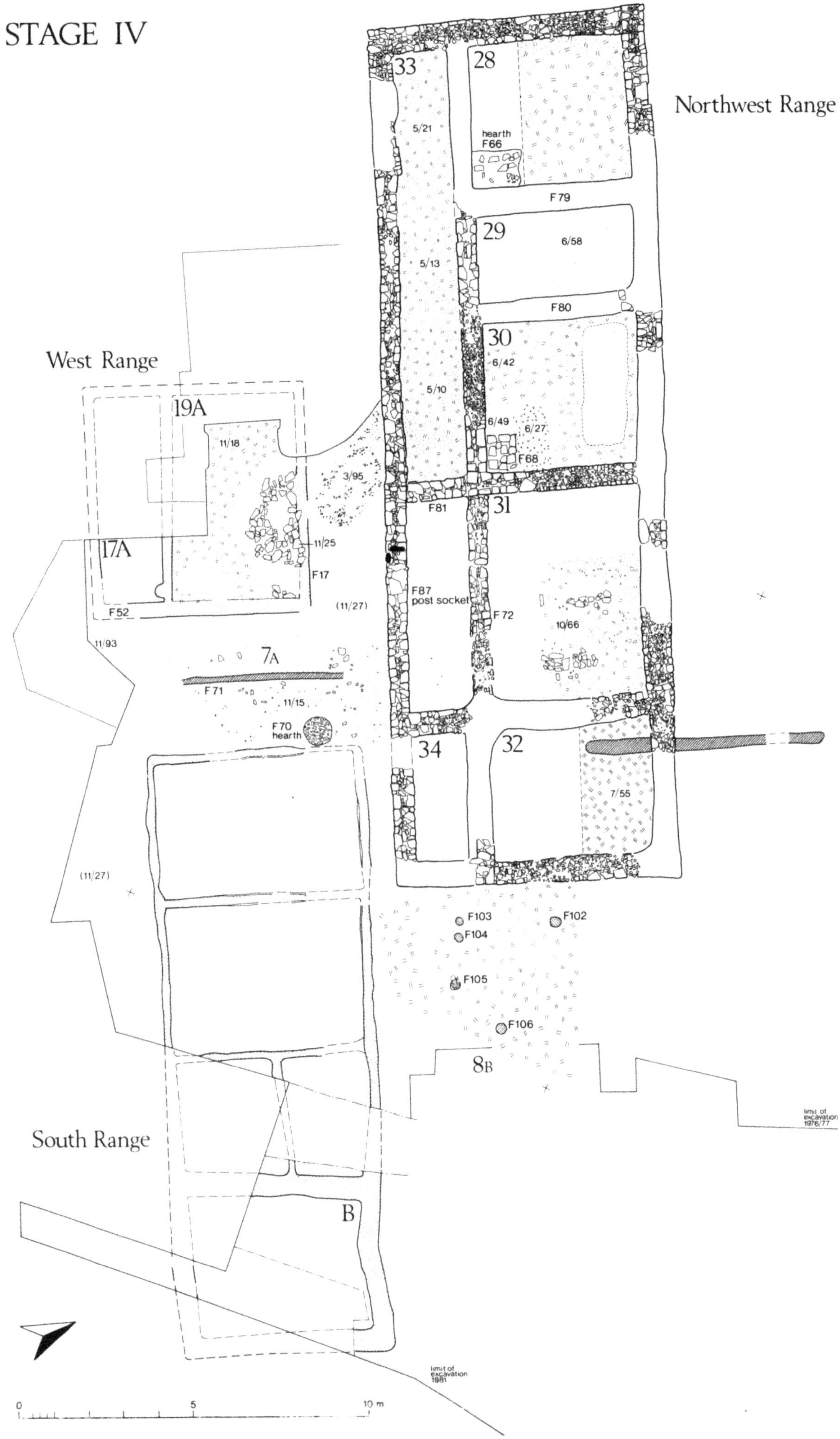
STAGE IV
Northwest Range
West Range
South Range
28
29
30
31
32
33
34
19A
17A
7A
8B
B
hearth F66
F79
F80
F81
F87 post socket
F72
F68
F17
F52
F71
F70 hearth
F102
F103
F104
F105
F106
5/21
5/13
5/10
6/58
6/42
6/49
6/27
10/66
7/55
11/18
11/25
3/95
(11/27)
11/93
11/15
limit of excavation 1976/77
limit of excavation 1981
0
5
10 m

debris beneath. In it were set two north-south lines of stones *c* 1.00 m apart. The purpose of which remains obscure. In the north-west corner of the room the clay demolition debris appeared to have been used directly as a floor surface. In the south-east the first patch of occupation refuse accumulated on top of a slightly larger spread of pebble, apparently a floor surface. This bore an ambiguous relationship with the foundation trench used when the north wall of the Corridor was extended across Room 31, possibly in Stage IV (p 26). Wall F72 had surfaces associated with it in the corridor to its south, a space left without evidence for flooring if that wall is not of Stage IV.

In Room 32, thin grey laminae of occupation debris lay above a flaky clay (7/55) which was either the top of the demolition debris or a surface in its own right.

In Corridor 33, clay make-ups 0.10 m to 0.15 m deep (5/10, 5/13, 5/21) formed a floor surface on top of which were two patches of sandy gravel, perhaps repairs to a minor subsidence into the pre-complex ditch. Cutting the bottom-half of the clay make-up but sealed by the top-half were two charcoal-filled pits 0.10 m to 0.15 m deep, demonstrating that the seemingly homogeneous clay had been deposited in at least two layers. The clay also sealed the foundations of two unused cross-walls matching the divisions between Rooms 28/29 and 29/30. These were not carried above floor level and were sealed across the full width of their foundation trenches by the clay, the eastern one having been backfilled after only two stones had been laid in the bottom.

Leaving aside the argument as to whether or not wall F72 belongs to Stage IV, Corridor 33 was not carried through the full length of the range, being interrupted opposite the eastern wall of Room 30 by butt-ended wall F81, which resembled the two unused wall foundations referred to in the previous paragraph. F81 survived *c* 0.05 m above the clay make-ups. At the eastern end of the range the alignment of Corridor 33 was resumed by a long narrow room (34) in which no coherent surface of Stage IV could be detected. Beneath the room, however was the semi-circular pool of Stages I-III (F27), backfilled with loam and blocks of Hawarden Sandstone identical to those used in the Stage IV construction. The top of the backfill lay at approximately the same level as the floor surface of Corridor 33.

South Range

An external surface (11/15), overlying the southern edge of the foundation trench for the North-west Range, abutted the base of the clay wall-stub on the west side of the South Range, that wall was still in use during Stage IV.

Later demolition debris immediately pre-dating Stage VII demonstrated that the South Range was standing till that time and, by inference, that it stood in Stage IV.

West Range (pl II, III)

It has already been argued that patching of the concrete floors in Room 19 and 20 implies the retention of these floors for some time during Stage III. Both floors were sealed by a clay deposit which also sealed the dividing wall between them. The northern parts of the clay lay beneath a later patching of sandstone slabs lying one-deep in loam (11/25), which clearly respected the northern wall-line; this patching was almost certainly a compensation for subsidence into the pre-complex ditch.

In Stage IV the West Range appears to have been reduced to one or perhaps two rooms only: Rooms 19 and 20 were amalgamated, to form Room 19A; late demolition debris covered the eastern part of the 17A/19A dividing wall, which may imply that it was not standing at the time the range was demolished. Not enough of this part of the range was excavated to be certain, but it is possible that the West Range stood as a single large room in Stage IV.

External Areas

The construction of the North-west Range effectively divided the large external space of Stage III into two (Yards 7A and 8B), connected by a narrow passage between the North-west and South Ranges (pl II).

Cutting surface 11/15 and covered by the subsequent demolition debris of the West Range prior to Stage V was gully F71. This sloped from its northerly point of origin to the south where it was cut by a later ditch, beyond which it did not reappear.

Against the outside of Room 11 and also cutting 11/15 was a shallow circular pit (F70) containing small pieces of sandstone set neatly on top of a brownish clay. The presence of charcoal and slight burning on the upper surfaces of some of the stones suggests that this might have been a hearth or the base for a brazier.

In Yard 8B was a group of small pits (F102-106) cutting either the demolition debris of the timber North Range or the charcoal waste from Stage III furnace F69. They were sealed by debris from the demolition of the South Range prior to Stage VII.

Near the eastern limit of excavation was a large charcoal-filled feature, probably a furnace. It is possible that the furnace was built during Stages I-III, in the absence of positive proof for an early dating, however, it is included here with Stage IV features post-dating the demolition of the timber North Range.

Figure 12: Plan of demolition between Stages IV and V, scale 1:150

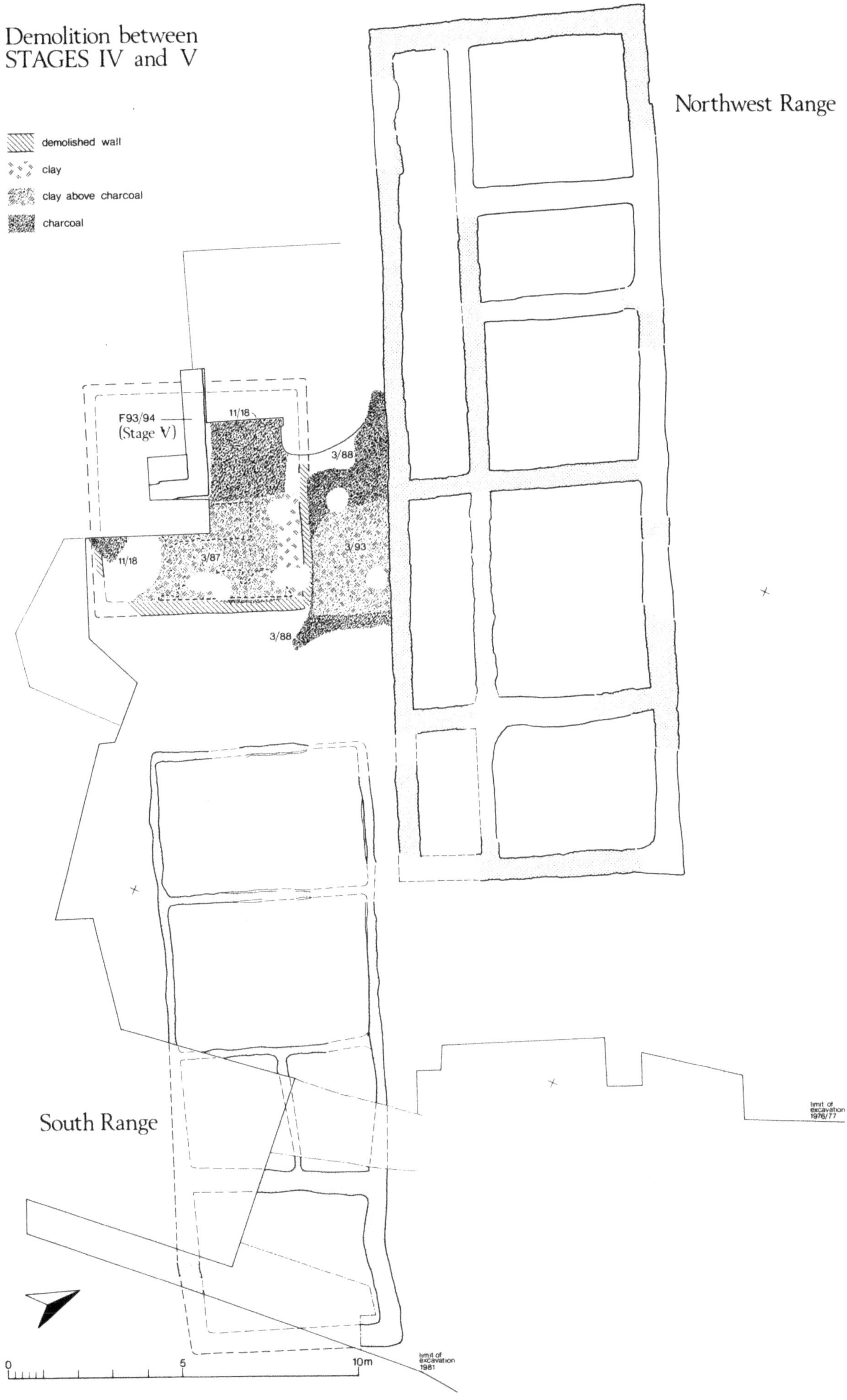
Demolition between
STAGES IV and V
demolished wall
clay
clay above charcoal
charcoal
Northwest Range
F93/94
(Stage V)
11/18
3/88
3/87
3/93
11/18
3/88
South Range
limit of excavation 1976/77
limit of excavation 1981
0
5
10m

Stratification possibly of Stage IV
North-west Range
At some time, probably but not demonstrably during Stage IV, the addition of wall F72 carried Corridor 33 the whole length of the range, apart from the south-east corner occupied by Room 34.

The new wall (F72) displayed the same method of construction as that characterising Stage IV and also used similar Hawarden Sandstone blocks. There were significant differences, however, in that its raft of igneous boulders was less closely packed and where its western end abutted wall F81, and the bottom of F72 was 0.10 m to 0.15 m higher.

The butt ends of F72, as well as its slightly different method of construction suggests it is later than the Stage IV walls of the North-west Range; and yet it resembles them more than anything else on the site. It is possible that Room 31 was originally built to span the full width of the range and was only subsequently reduced in size to allow Corridor 33 to run through to the east. Alternatively, F72 may simply have been the last wall in the range to be built.

The first layer to abut wall F72 in Corridor 33 was a pebbly gravel and loam (5/5). This also abutted both sides of F81 and ran the length of Corridor 33. Similar material (7/71-7/73) had been used as patching in Room 31. If these layers were contemporary with the construction of F72 and that wall was part of Stage IV, then the clay layers 5/10 etc illustrated on Fig 11 are more likely to have been Stage IV make-up than surfaces in their own right. On the other hand it is quite possible that 5/5 was a refurbishing of the surface later during Stage IV, and was retained in Stage V.

Dating evidence for Stage IV (PVW)
This stage includes some Antonine pieces (nos 80, 85, 90) and a number which are either late Hadrianic or Antonine (nos 86, 94). A late second century date is likely.

Demolition between Stages IV and V
A sequence of deposits similar to those marking the demolition between Stages III and IV represented the demolition of the remaining parts of the West Range prior to, and possibly in preparation for, the construction work representing Stage V. Minor alterations were also made in the North-west Range (Fig 12).

West Range
Clean powdery charcoal (11/18), between 0.02 m and 0.04 m deep, covered the Stage IV floor surfaces of the West Range. On top of the charcoal was a partially articulated crumbly white wall-plaster, covered in turn by brownish-orange clay partially burnt red; *c* 0.10 m deep, this contained no sign of post or wattle impressions. The charcoal, uncontaminated by domestic finds, nails or roofing tiles, suggests the burning of a thatched or shingled roof; the reddened clay testifies that parts of the walls, at least, were affected by the fire. The stratification of the plaster, as articulated sheets beneath the remnants of clay walling, implies that, in places, whole sections of walling collapsed during the fire or were subsequently pushed over, rather than being hammered to the ground in smaller pieces.

A hollow in the top of the demolition debris, above the line of the pre-complex ditch was levelled with clayey loam. An earlier patching (11/25 in Stage IV), similarly located above the earlier ditch, showed that subsidence had also occurred while the West Range was still standing; it may even have accelerated the decay of that range.

STAGE V

A hypocausted room was constructed above the debris of the West Range and the three rooms of the North Range were rebuilt in stone (Fig 13 and pl XI, XII).

In all cases, foundation trenches for the new walls clearly cut stratification of Stage IV or, in the case of the hypocausted room, the demolition deposits overlying them. The hypocausted room lay beneath the Stage VI Bath-house and the rooms of the new North Range were stratigraphically earlier than the demolition of the South Range, which itself preceded the construction work of Stage VII. Rebuilding of the North Range may therefore be placed in Stage V or Stage VI. It is described here as part of Stage V because it shares with the hypocausted room the use of distinctive green sandstone blocks, not seen in the building work securely attributed to Stage VI.

The remains of the new North Range were fully excavated but time permitted only limited excavation in the hypocausted room, to establish the general character of its walls and its place in the sequence.

Hypocausted Room
The extent of the new room (37) is virtually unknown, the excavated walls defining only its north-eastern corner. The area further to the south was not investigated, while that to the north-west had been cut away by medieval robbing.

Only two wall alignments were uncovered (F93, F94), both founded in trenches 0.65 m wide and cut to a depth of 0.30 m from the surface of the demolition debris of the South Range, through the underlying concrete floors and into natural subsoil. Two courses of wall footing remained, 0.57 m wide, comprising long thin pieces of greenish sandstone laid mainly as

Figure 13: Plan of Stage V, scale 1:150

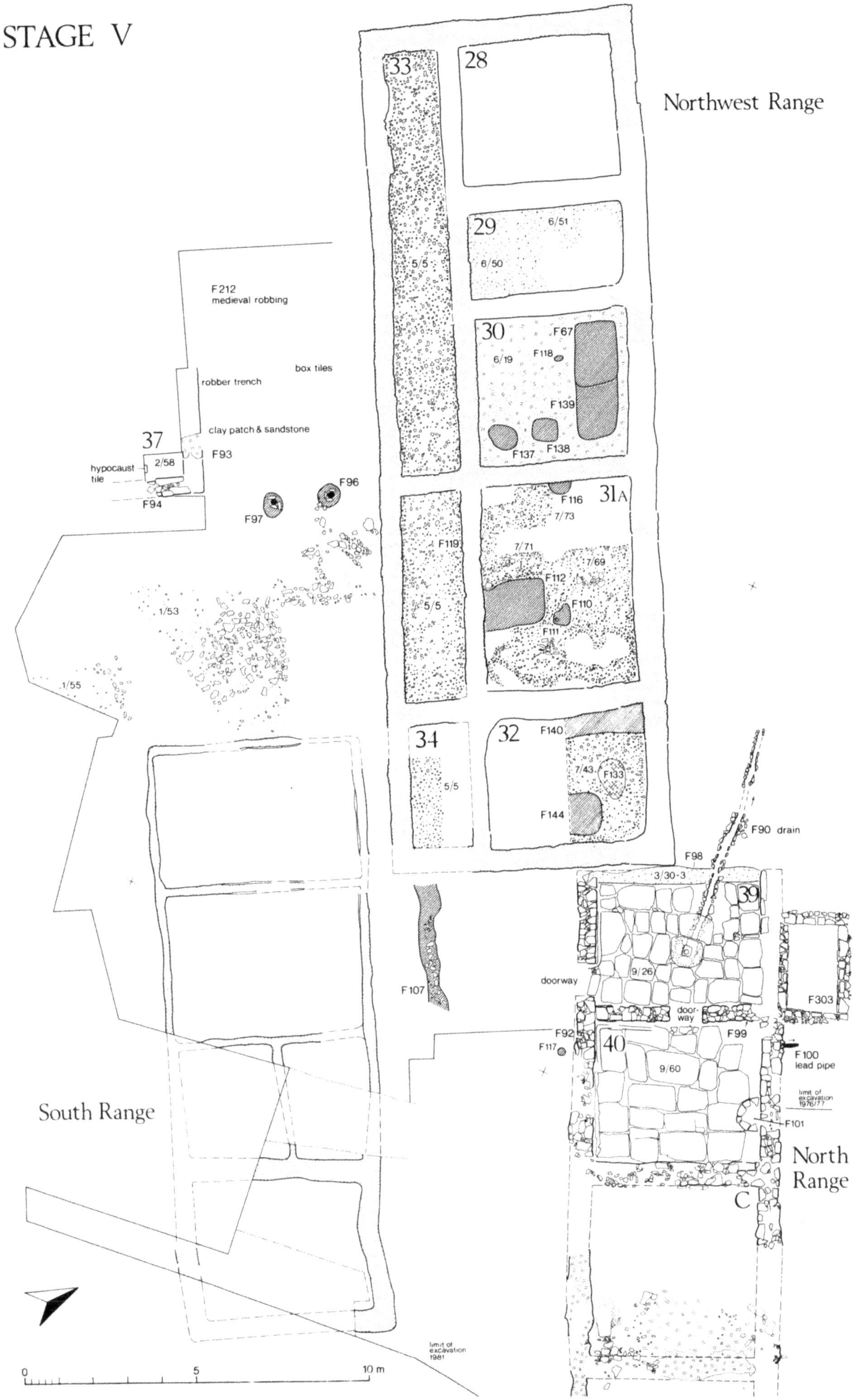

STAGE V
Northwest Range
33
28
29
6/51
5/5
6/50
F212
medieval robbing
30
F67
F118
6/19
box tiles
robber trench
F139
clay patch & sandstone
37
F93
F137
F138
hypocaust tile
2/58
F96
F116
31A
F94
F97
7/73
F119
7/71
7/69
F112
1/53
5/5
F110
F111
1/55
34
32
F140
7/43
F133
5/5
F144
F90 drain
F98
3/30-3
39
9/26
doorway
F107
door-way
F303
F92
F117
40
F99
F100
lead pipe
9/60
limit of excavation 1976/77
South Range
F101
North Range
C
limit of excavation 1981
0
5
10 m

stretchers. The upper parts of the walls were robbed before the building of the Stage VI Bath-house.

Enclosed within these walls was a mortar floor on which lay a single hypocaust tile. Covering the mortar floor and contiguous with the edges of the tile was a layer (0.05 m deep) of dense charcoal, proving that the system had been fired. These deposits were sealed by the make-ups for the Stage VI Bath-house sub-floors.

North Range

The walls of the new North Range were 0.5 m to 0.7 m wide. The foundations were irregular, but generally shallow (*c* 0.15 m). The east wall, and all east-west walls, were built on edge-set pieces of sandstone packed one course deep, often in a dense pink clay.

The foundation for the north wall of Room 39 was reinforced where it bordered pit F303; here, it employed a crude herring-bone technique, with a mixture of yellow and purple sandstohe set in a matrix of clays in a foundation trench 0.55 m wide and at least 0.40 m deep.

The east wall of the range apart, the north-south walls had no real foundation, the bottom course of the walls between Rooms 39 and 40, and between Rooms 40 and 41, being slightly impressed below contemporary ground level into the natural subsoil.

The western wall of Room 39 had foundations which consisted of unmortared pieces of yellow and purple sandstone with some glacial boulders, generally standing free but sometimes in a very shallow foundation trench (0.05 m deep maximum), and varying in width between 0.45 m and 0.60 m.

The walls above displayed a rough uniformity, characterised by the use of purple micritic Bunter Sandstone, the nearest source for which is the Wirral. The facings laid variously as headers and stretchers, sometimes occupied the full width of the wall, but in other positions left room for a narrow central core of smaller pieces of sandstone. The average width of the external walls was 0.55 m, while internal partitions were slightly narrower.

Breaks in the walling and plaster at the south-east corner of Room 39 and between Rooms 39 and 40 marked the positions of doorways. At various points around Rooms 39 and 40 were short lengths of coarse *opus signinum* plaster, poorly-preserved but almost certainly the remnant of a quarter-round moulding forming a skirting round the rooms. Nowhere was the plaster flush with the wall and panelling or a thick plaster finish (subsequently decayed or removed) must be invoked to explain the fairly large intervening gap (0.05 m to 0.10 m), found filled with brown loam.

Overall the new North Range measured 14.35 m east-west (not including the fourth room extending east) by 6.0 m north-south (to the outer faces of walls). Rooms 39 and 40 were approximately the same size (4.8 m north-south and respectively 4.3 m and 4.0 m east-west between the inner faces of the walls). The easternmost of the three new rooms (C), like its timber predecessor, was larger than other rooms in the range, though not in this case open-sided. Internally, it measured 4.8 m north-south by 5.9 m east-west. A further room lay to the east of the excavated area.

Both Rooms 39 and 40 were floored with large flagstones. The flagstones in Room 39 (9/26) were of purple Bunter Sandstone 0.05 m to 0.12 m thick, bedded in up to 0.07 m of yellow sand, in Room 40 (9/26) they were a more even yellow-brown Hawarden Sandstone, 0.05 m to 0.15 m thick. Some of the central flagstones in Room 39 showed signs of having been disturbed. Drain F90 extended from the centre of the room beneath the stone floor (pl XVI).

In the north-west corner of Room 40 a small area was filled to the same level as the flags with whitish gravel which surrounded a lead pipe (F100) passing through the north wall and draining into a ditch F304) to the north of the complex (Fig 19). The gravel might be seen as a repair after the removal of a length of piping formerly connected with F100. In the east, resting on top of the flagstone floor, a semi-circular arrangement of stones (F101) may have been a hearth or a support for a basin. Evidence for the flooring of Room C at the eastern end of the range had been removed by levelling down in preparation for the building of a twentieth century pig-sty. This was demolished in the 1950s but the present occupant of Pentre Farm recalls that the divisions within the pig-sty were formed by slabs of stone similar to those that floored Room 40 of the Roman building.

After the construction of the range but before the Stage VII resurfacing, the floor of Room 39 had subsided into the fill of the Stage I clay-lined pit F4. Primarily in this hollow, but also on other parts of the floor, occupation refuse had accumulated in the form of compacted brownish loam containing several finds.

External Areas

In Yard 8B linear feature F107, whatever its function, was presumably still in use because the debris of the South Range demolished prior to Stage VII, obeyed its line.

A shallow trench (F107) cutting and sealed by the same deposits as the pits, lay to the west of Yard 8B parallel with and 1.5 m away from the wall of the South Range. Demolition debris of the South Range prior to Stage VII obeyed the line of F107, which presumably remained in use until that time. Its function is unclear, but may represent the foundation of a lean-to structure.

Stratification possibly of Stage V

External Areas.

In the north-west corner of Yard 7A, immediately above the demolition debris of the timber West Range was a gravelly loam matrix with scattered flat-laid pieces of sandstone (1/53). The top of the demolition debris was uneven, but showed no sign of silting in the surface depressions; presumably layer 1/53 soon sealed it. The latter was cut by post-pits related to Stage VII.

Isolated deposits in the North-west Range Stage V
The earliest layers of this range have already been described in Stage IV. Those described here represent the next general phase, but cannot be linked with certainty to features attributed to Stage V.

Stratification in Room 28 was not recorded. In Room 29 two contiguous clayey deposits (6/50, 6/51) formed an unbroken floor which lay at the same absolute level as the surface of a floor of clayey-loam (0.10 m to 0.15 m deep) in Room 30 (6/19). The apparent division in the surface of Room 29 was probably caused by slight differential settlement of the floor onto features of earlier timber structures.

In Room 30, floor 6/19 was cut by a long shallow pit (at most 0.20 m deep) F139, re-cut before the laying of the surface described in Stage VI by an equally regular, but shorter and shallower rectangular pit F67. On the eastern side of the room were two smaller sub-rectangular pits (F137, F138), both too shallow to penetrate the full depth of floor 6/19. The base of pit F138 was paved with randomly-placed pieces of sand-stone. Running diagonally across the room from south-east to north-west was a depression in the floor surface which did not appear to be a cut feature. Its fill of brown loam may have been deposited as part of the levelling process prior to the laying of the floor (6/13) described in Stage VI. In the side of the depression a stakehole had been formed (F118).

In Room 31A a floor of compacted pebbly loam (7/69) resembled the surfaces in the south-west corner of the rooms (7/71-73) and in Corridor 33 (5/5), perhaps suggesting contemporaneity. Floor 7/69 etc was cut by a small pit (F110, itself cut by stakehole F111); it lay adjacent to a much larger pit, F112, with irregularly sloping sides. Beyond this floor in the east was another small pit (F116). Above the floor was a patch of gritty loam cut by two small pits.

Room 32 was reinforced with a pebbly loam (7/43) similar in composition to and at the same absolute level as the flooring just described for Room 31A. Floor 7/43 was sealed by a series of charcoal deposits, the earliest of which filled a possible hearth (F133) in the middle of the room. Interleaved with the charcoal were lenses of pebbly mortar, beaten clay and lead slag These deposits were cut by two pits filled with building debris. Two pits (F140, F141) were filled with a loam completely free of charcoal and these may therefore be robber pits for installations in what appears to have been a workshop. In the north-west corner of the room was a carved stone bench-footing (Fig 34, no 4). Its base had been set 0.02 m below the level of the earliest identifiable floor of the room but some charcoal deposits had lapped up against it, proving that the bench must have been present during the deposition of the charcoal. The general character of the charcoal and its lenses closely resembled that of the floor in Room 41.

A small area of pebbles remaining on the south side of Room 34 has been treated as the same as the material resurfacing Corridor 33 (5/5).

Dating evidence for Stage V (PVW)
Colour coated white wares (from Cologne or the Nene valley) (no 136) first appear in this Stage. Nos 109 and 112 are both likely to date from the second half of the century or early third century.

STAGE VI

Stage VI consists of a Bath-house, built where the hypocausted room (Stage V) had stood (Fig 14 and pl V-IX). The northern and eastern parts of the Bath-house were excavated in detail, though lack of time prevented full investigations of the parts nearest the southern limit of excavation, where only the upper-most stratification was dug. In the absence of evidence to the contrary, the north, north-west and southern ranges are assumed to have remained standing. A room (41) added to the north east corner could have been constructed in either Stage V or Stage VI but is described here for the sake of convenience.

The only surviving wall excavated in detail (F122) was set in a shallow trench half filled with clay-packing, beneath a footing of dressed sandstone blocks (0.70 m wide). Above this the wall, inset by 0.05 m in each side, survived as up to five courses of poorly-dressed sandstone blocks, crudely arranged as headers and stretchers. Between the facings was an infill of clay and small pieces of sandstone.

About 1.5 m to the east of the Bath-house, Verandah 38 was defined by a number of pits (F123-126) between 2.4 and 2.6 m apart, varying between 0.5 and 0.7 m in depth.

Bath-house
All of the Bath-house walls had been robbed, apart from those around Room 47. Wall alignments were difficult to detect as a consequence, and could only be inferred from the remains of other features. Altogether four rooms (47, 48, 50, 51), a verandah (38) and two corridors (49, 51A) could be defined in this way. The *praefurnium* presumably lay on the south wall of Rooms 47 or 48 or on the west side of Room 48.

Room 47 was formed by three extent walls and a small part of a fourth. The southern and eastern boundaries of Room 48 lay beyond the limit of excavation, but have been assumed to continue the established lines of the adjacent rooms. That the northern boundary of Room 48 was a continuation of that of Room 47 was shown by the edge of a hypocaust sub-floor, contiguous with but not cut by a medieval robber pit. The eastern and western walls of Room 51 were implied by the straight edges of its concrete floor (2/28) as well as by the fragmentary remains of clay and timber walls (3/81). One stone remaining in the middle of the projected western alignment of Room 51 need not be an indication of wall-width, as it may have been a threshold block.

The south and east boundaries of Room 50 have already been provided by the outlines of Rooms 48 and 51. The western alignment partially survived in a short length of walling, 0.7 m wide, (F144) abutting the western side of plunge bath F130. Another piece of walling 1.5 m long and at least 0.7 m wide, indicated the northern limit of the room (F145). Contiguous with this wall to the north was the main Bath-house drain, much of which had been destroyed by medieval robbing. The drain appears to have originated in the centre of Room 51 draining out of the northern side of the room where it turned sharply to the west, running alongside Rooms 49 and 50. Its easternmost extant walls, just to the north-west of Room 51 were constructed of box-tiles. Nowhere did these survive above one course but, given floor levels in Room 51 it is possible that they represent the full depth of the drain at this part (*c* 0.30 m to 0.35 m). Further west, however, the drain became appreciably deeper, its wall being formed by large sandstone flagstones up to 1 m long and 0.80 m to 1 m deep.

At the part where this flagstone lining survived best the opposing flagstones leaned towards one another: at the bottom, the drain was *c* 0.40 m wide but the gap between the tops of the flagstone lining was a mere 0.20 m to 0.25 m. Here the drain was capped by a few Bunter Sandstone blocks one course deep which aligned with one course of similar stones just over a metre to the south (F134). Taken together these may represent the ruined outer face of a wall or tentatively a kerbing. A compacted pebble surface (2/20), which slightly overlapped the southern side wall of the main drain (F95) appeared to represent the alignment of stones in F134. The pebble surface was cut by gully F204 of Stage VIII, so its position in Stage VI as opposed to Stage VII is not stratigraphically secure.

East of the Bath-house, interpretation of post-pits F123-126 as the foundations for an open verandah was confirmed by the subsequent accumulation of gravel layers which while interleaving between the post-pits were generally confined either to the verandah itself or to the open Yard to its east. Within the verandah the gravel layers abutted the eastern face of the Bath-house, which was therefore in existence before they were laid. The post-pits themselves had been dug through the gravel surfaces but because the easternmost pit (F126) was cut by trench F154 of Stage VII the verandah as a whole has therefore been phased independently in Stage VI. The first floor or make-up to postdate the eastern wall of the Bath-house wall was a layer of mixed clay and grit (3/38).

In Room 47 charcoal and mortar overlying wall F93 of the Stage V hypocausted room seemed to be the first of four make-ups for a flagstone and baked clay sub-floor (2/25), sealed by the Stage VII hypocaust stacks.

In Room 48, standing on a clayey loam sub-floor founded on crushed tile and mortar, was a hypocaust system, the height of which could be gauged at 0.75 m, from a fallen but articulated *pilae* near the east end. The stacks were usually 0.45 m to 0.55 m apart, centre-to-centre.

Room 50 had been completely destroyed by medieval robbing except for a plunge bath (F130) with a lead drainpipe (F131) leading from a plug-hole in its north-east corner to the main Bath-house drain to the north. The trench in which the pipe lay was filled with several well-compacted tip-layers; it had eventually become choked with clayey silt.

In Room 51 a well-preserved floor of large ceramic tiles (2/23) was founded on a stone and mortar make-up, and a series of firmly compacted levelling layers which partially filled the extreme eastern end of the construction trench for the main drain. Strips along the eastern and western side of the tiles were filled by *opus signinum* (2/28) which marginally overlapped the edges of some of the tiles, including the one that crossed the line of the wall dividing Room 51 from Room 47. This was clearly a threshold and had been heavily scored by repeated opening of the door that once stood in this position.

The whole floor was laid so as to drain towards the centre and to the north, to feed into the main drain by way of an inlet still represented by a pink mortar surround, *c* 0.10 m square. A layer of clay and plaster occurring in Verandah 38 and by the main drain in Corridor 51A indicated a possible route of access at that point (3/80).

North, North-west and South Ranges

All three ranges are assumed to have remained in use. Stage VII alterations to the North-west Range and Stage VIII reflooring of the North Range imply that both stood in Stage VI. Similarly, deposits in Yard 7, bounded by Verandah 38 (added in Stage VI) ran up to, but not beyond, the western wall of the South Range, demonstrating that at least that part of the range stood in Stage VI.

At about this time the doorway between Rooms 39 and 40 in the North Range was blocked with pieces of stone (F156) like those used in the original construction of the Range, but more irregularly laid.

Stratification possibly of Stage VI

Room 41

In the angle between the North and North-west Ranges, Room 41 was built on unmortared footings of varying width, sometimes in a very shallow trench (0.05 m max) but generally free-standing. This, together with the absence of mortar and the lack of proper coursing,

Figure 14: Plan of Stage VI, scale 1:150

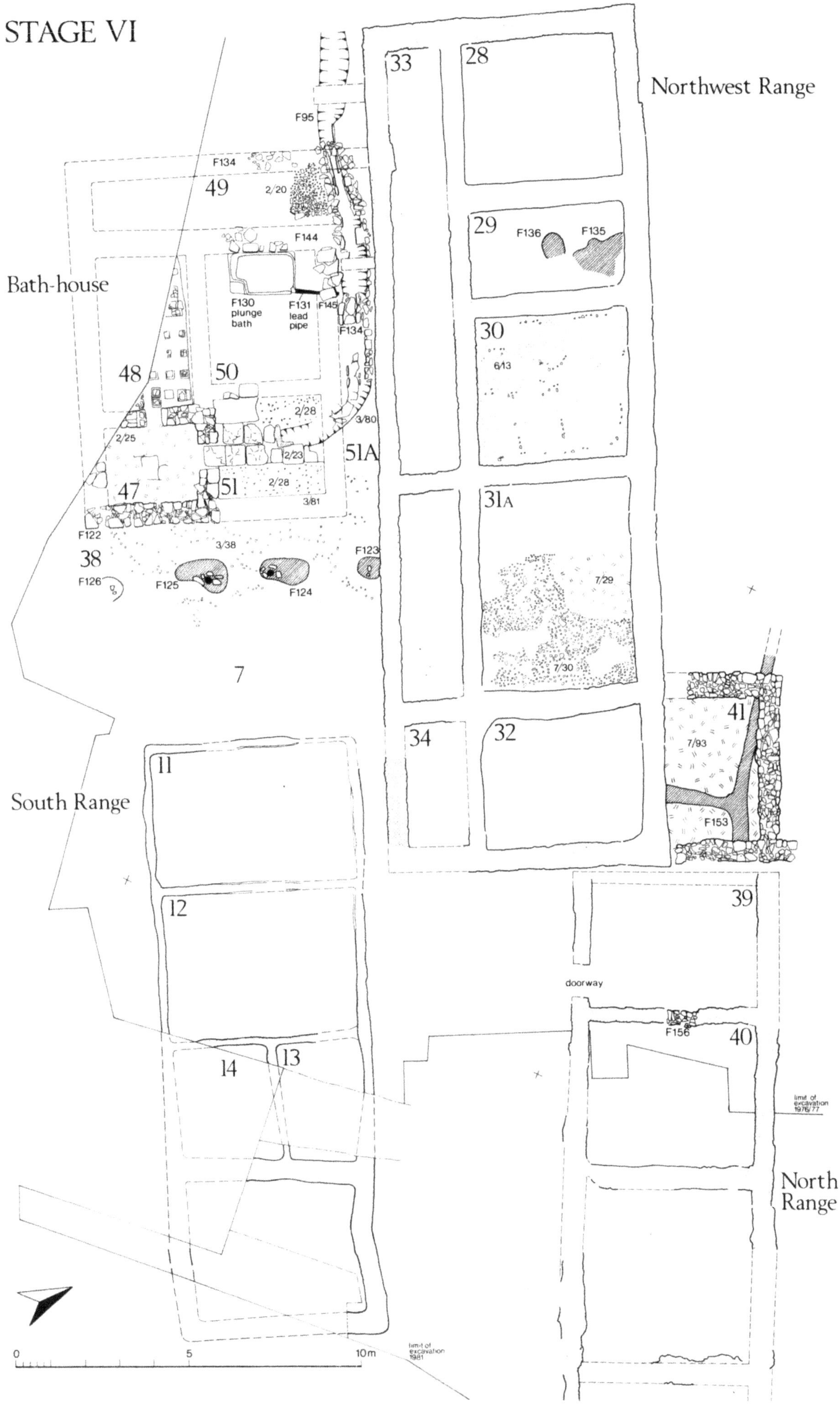
STAGE VI
Northwest Range
Bath-house
South Range
North Range
33
28
29
30
31A
32
34
F95
F134
49
2/20
F144
F136
F135
F130 plunge bath
F131 lead pipe
F145
F134
48
50
6/13
2/28
3/80
2/25
2/23
51A
47
51
2/28
3/81
F122
3/38
F123
38
F126
F125
F124
7/29
7/30
7
41
7/93
F153
11
12
39
doorway
F156
40
14
13
limit of excavation 1976/77
limit of excavation 1981
0
5
10m

implies a half-timbered superstructure.

These footings formed the north, south and east sides of the room, the southern side being formed by the north wall of Room 32. Since the eastern boundary of Room 41 duplicated the western wall of Room 39 the new room should be seen as an addition to the North-west Range, rather than to the north.

The northern and eastern footings overlay the western part of Stage V drain F90. Occupation deposits within the room were cut by gully F153 running to the north of the complex via gully F306 into ditch F317, which had filled up with charcoal detritus during Stage VII. On the assumption that gullies F153/F306 would not have lead to a choked drain, Room 41 belongs to Stage V or Stage VI (Stage VI is preferred here for the sake of convenience).

Within Room 41 a clay floor was covered by fine laminae of lead oxide and charcoal, interleaved with patches of clay. The lead oxide could have been used as flooring material or make-up and need not imply lead-working. Almost identical stratification occurred at the same absolute level in Room 32 to the south, where laminae of charcoal and lead oxides interleaved with the charcoal fill of a shallow three-phase hearth (F133, Stage V) in the middle of the room.

Isolated deposits in the North-west Range, Stage VI

Some or all of the layers described in this section may have belonged to Stage VI, though they lack clear stratigraphical relationships to demonstrate the point. For convenience, they are illustrated on the Stage VI plan (Fig 14).

In the north of Room 29, cutting clay floor 6/51 of Stage V were two pits (F135, F136) truncated by a Stage VII pit (F159).

In Room 30, above the floor attributed tentatively to Stage V (6/19) was a possible floor of compacted sandy loam (6/13).

Room 31A was floored by a compacted clay-loam (7/29-30) above which was a patch of sandy-loam.

Corridor 33 appeared to remain unaltered from Stage V. In Corridor 34 all stratification later than 5/5 (Stage IV) was removed without examination.

Demolition between Stages VI and VII

A sequence of layers covering Rooms 11 and 12 implied demolition of the South Range; evidence for the eastern parts of the range had been removed by modern levelling. The eastern end of Corridor 33 in the North-west Range was also altered at this time.

South Range

Many of the post-positions in the walls surrounding Rooms 11 and 12 were filled with homogenous brown loams, presumably the result of slow decay after the posts had been cut off. Partially-articulated sheets of white wall-plaster lay face-down on the concrete floors of Rooms 11 and 12.

Above the plaster sheets and sealing the post-positions along the wall-lines were several layers of clean, dense, reddish-brown and orange clay, up to 0.25 m thick and often mixed with small lumps of wall-plaster. Trapped between the clays at one point were two straight, thin (*c* 0.01 m) bands of brown loam, up to 2 m long, sloping at a shallow angle to the horizontal. Similar bands in the stubs of the walls themselves have already been interpreted as wattle impressions (p 9). If the same interpretation holds true here, it appears that some parts of the walling had been pushed over in sections, and not been broken down piecemeal.

Above Rooms 13 and 14 the clay was thinner (0.15 m) and neither contained nor sealed any plaster.

The stratigraphic sequence suggests that the timber South Range was demolished while the stone North-west Range stood intact. The subsequent modification to the south-east corner of North-west Range must have occurred very soon after the demolition of the South Range because all the stratification capping the demolition debris related in one way or another to the new walls of Rooms 42 and 45 (Stage VII).

Dating evidence for Stage VI (PVW)

The latest pieces from this stage are nos 138, of late second to early third century (from a construction or immediately post-construction layer) and 147 and 150 (from occupation levels), both of mid to late second century. The developed type of dish with intersecting arc decoration does not seem to be present in Antonine Scotland and a date of introduction prior to *c* AD 175 seems improbable. Stage VI should, therefore, post-date the late second or early third century.

STAGE VII

The levelled clay walling of the demolished South Range was superseded by Rooms 42 and 45. To the east of Room 45, above former Yard 8 were two possible sheds or enclosures (43,44). South of Rooms 42 and 45, an outdoor area was laid (1/33) above the demolished South Range. In the Bath-house, alterations were made to Verandah 38 and to Rooms 47 and 51 (Fig 15). All rooms were fully excavated apart from Rooms 45, which was half-sectioned and then reduced, without record, to the level of subsoil late in the excavation (pl IV). In the eastern parts of the site, evidence of Stage VII had been removed, partially by ploughing but mainly by levelling for modern farm buildings.

Figure 15: Plan of Stage VII, scale 1:150

STAGE VII

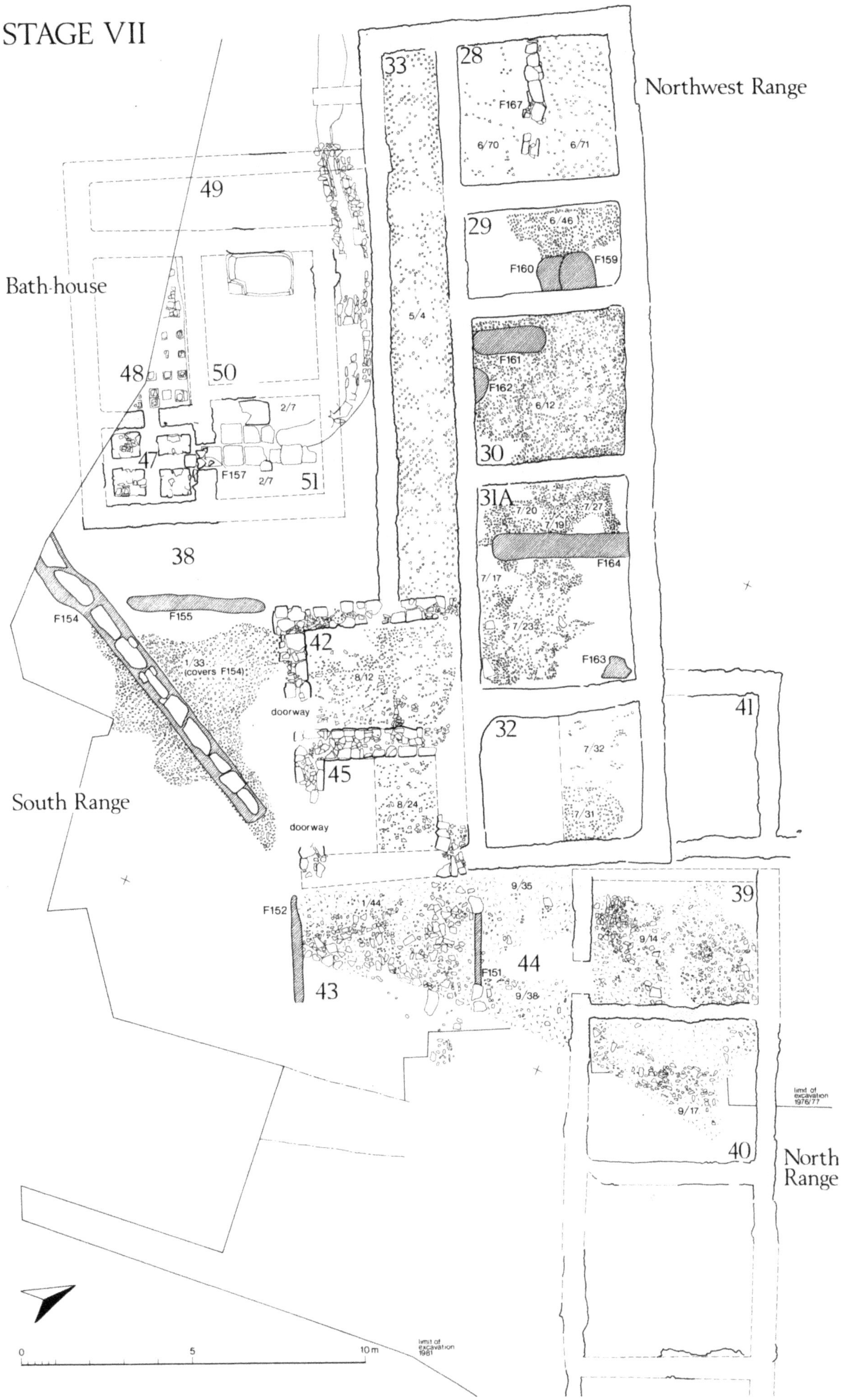

Northwest Range
Bath-house
South Range
North Range
doorway
doorway
limit of excavation 1976/77
limit of excavation 1981
0
5
10 m

Rooms 42 and 45

The foundation trenches for these rooms cut the clay demolition debris of the South Range. The surviving walls of these rooms were not bonded together but the general similarity of building materials, method of construction and level of earliest associated surfaces suggests they were constructed together. The earliest stratification post-dating the construction of these walls was sealed by layers equated with the resurfacing of Yard 46 in Stage VIII.

Where the concrete floor of the former South Range had been used as a foundation, the foundations had left shallow impressions a centimetre or so deep. Where the concrete was absent, foundation trenches were dug, about 0.95 m wide and up to 0.50 m deep from the surface of the demolished clay walling. Within these foundation trenches were set three courses of sandstone blocks laid as paired stretchers, alternating with two layers of dense orange clay. Above these foundations, survived up to two courses of walling. This was made of poorly-dressed blocks up to 0.45 x 0.25 x 0.15 m, set as stretchers with an infill of smaller pieces of sandstone, loam and off-white mortar. In Room 42 walls survived on the southern and western sides measuring between 0.60 m and 0.70 m in width. In Room 45, where the surviving walls were wider (0.75 m to 0.90 m) and also bonded together in the north-eastern and south-western corners. Enough of the northern wall of Room 45 survived to show that it once abutted and duplicated the south wall of Room 32. In contrast the south wall of Room 31A was clearly used as the northern wall for Room 42. Breaks in the walling in the south-east corners of both rooms (filled by shallow robber trenches) marked the positions of doorways.

In preparation for the laying of floors the demolition debris beneath Rooms 42 and 45 was levelled off with a variety of clay, gravels and mixed rubble. These were covered by surfaces of glacial pebbles and tile fragments pressed into a clay/loam matrix (8/12, 8/24); similar surfaces were found in all eastern and south-eastern parts of the complex.

North Range

Sealing the Stage V flagstone floors and later than the abutting blocked doorway between Rooms 39 and 40, were two new floors of clay with pebbles, tile fragments and pieces of sandstone (9/14, 9/17). Their surfaces were horizontal, although the layers as a whole varied in depth between 0.05 m and 0.20 m. They closely resembled most other floors attributed to Stage VII, in the North-west Range and in Room 42 and 45.

Cutting the middle of the floor in Room 39 was an east-west row of thirteen irregularly-shaped stake-holes, interpreted as a wattle partition.

Structures south of the North Range

Three clay deposits (1/44, 9/35, 9/38) merged with one another to form a continuous make-up for surfaces of not only pebble and tile typical of Stage VII, but also sandstone slabs more confined to the limits of 1/44. These slabs appeared to have been bounded on the north by a slot (F151) which, although cut into the clay may still have been part of the same building walls. The loam filled slot began 0.70 m from the east wall of Room 32, beneath a large flat slab of sandstone and ended similarly, 2.85 m further east. Beneath both stones the slot narrowed in section from 0.10 m square into tenon-like projections 0.03 m wide, 0.10 m deep and 0.10 m long. If the excavated slot represents a beam, the mortices are difficult to explain unless the beam was being re-used. The stones themselves may have been post-pads for a lean-to.

The concentration of sandstone slabs above clay 1/44 was delimited on the south by another loam-filled slot (F152) apparently also cutting the contiguous clays but of a different character, being 0.05 m wider, less square in profile and without post-pads. It was traced from just outside the south-west corner of Room 45, for a distance of 3 m before being cut by modern disturbance. The two slots and surface 1/44 may represent a lean-to structure on the east end of the North-west Range.

External Areas

South of Rooms 42 and 45, the first feature to cut the demolished clay walling of the South Range was a small hearth (F158) lined with orange clay and filled with a gritty white lead waste lying beneath some charcoal. The lead waste contained a high percentage of lead oxide (identified by Dr Alan Williams of UMIST) but the hearth cannot be proven conclusively to have been used for the cupellation of lead.

Cutting the hearth was what appears to have been an abortive drain (F154), a vertically-sided trench sloping downwards from a depth of 1.0 m at its rounded point of origin outside Room 45, to 1.5 m where it passed beyond the limit of excavation, close to the south-eastern corner of the Bath-house. The base of the feature was filled to a depth of *c* 0.30 m with dense red-brown clay capped by uneven roughly rectangular slabs of Hawarden Sandstone resting 0.05 m to 0.20 m apart and covering the central parts of the trench only. Above these the trench had been refilled with a gravelly loam, the upper parts of which merged with the gravelly loam (1/33) sealing the feature.

Bath-house

Four crudely-built piers in Room 47 included one which partially consisted of *pilae* lying on their sides, still articulated by their original mortar (pl VII). Such *pilae* were first used in Stage VI and may have formed the original hypocaust system in Room 47. On p 36 arguments are advanced to show that the Bath-house was not used as such in Stage VIII; the new piers must be roughly contemporary with the rebuilding of Stage VII. Apart from the *pilae*, the new piers consisted of a

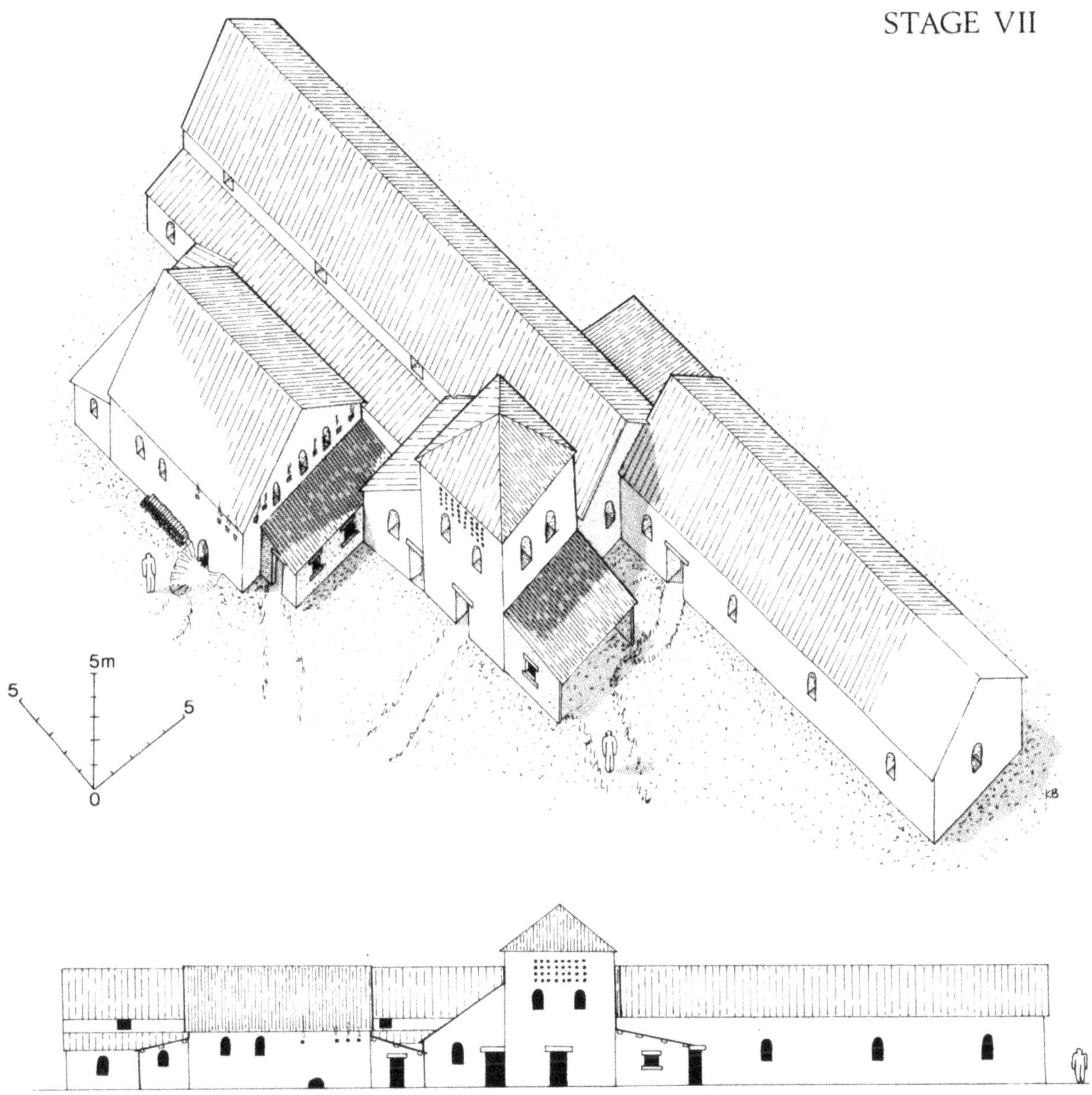

Figure 16: Tentative reconstruction of Stage VII

mortar and stone conglomerate, and measured 0.80 m to 0.95 m north-south, by 0.55 m to 0.80 m east-west. In the middle of the northern side of the room one square stone, perhaps the remains of flooring, lay flush with the top of the northernmost piers on a small piece of mortar conglomerate dumped at the end of the flue. To the north of the stone, the former doorway between Rooms 47 and 51 had been blocked with stones (F157), set in the same type of mortar conglomerate. In Room 51 two stone slabs (2/7) resting on the remains of a crude mortar portal above the Stage VI floor may represent the remains of largely robbed reflooring, possibly contemporary with the blocking of the doorway to Room 47. The later disturbance of Room 49 made it impossible to detect comparable changes there and Room 48 showed no signs of alteration.

One other feature may relate to this phase. Cutting some of the gravels at the eastern side of Verandah 38 (described in Stage VI above) was a shallow, concave north-south cut (F155) 0.05 m to 0.10 m deep and 3.90 m in length. Both ends were rounded, the northern end lying immediately to the south of the south-west corner of Room 42. This may represent the foundation of a re-built verandah - in part making use of the west wall of Room 42.

Isolated deposits in the North-west Range, Stage VII

Layers illustrated on the plan of Stage VII (Fig 15) are similar surfaces lying at similar absolute levels, which through the relationships of the surface (5/4) in corridor 33 are thought to immediately post-date the construction of Rooms 42 and 45. Features on the plan are those which through their regularity may have been dug for structural purposes. After removal of ploughsoil the extant stratification in Room 28 comprised two layers of greyish-brown charcoally loam (6/70, 6/71) separated by an east-west wall (F167) made of irregularly laid slabs of Bunter Sandstone one course deep.

In Room 29 a compacted floor of clay-loam with some pebbles (6/46), resembling the Stage VII floor in Corridor 33, overlay pits F135 and F136, described in Stage VI. The later features in Room 29 were two pits (F159 and F160) which were, in turn, cut by a pit that was thought to post-date the demolition of the complex.

In Room 30 a pebble-loam floor (6/12), similar to that in Room 29 and Corridor 33 overlay the stratification described in Stage VI. Two pits (F161, F162) cut 6/12 in the south-western quarter of the room. A layer of clay in the southern part of the room may have been laid as a repair.

Like floors in other rooms ascribed to Stage VII, Room 31A also had compacted pebble and tile surfaces set in brown loam (7/19, 7/20, 7/23). In the north-east corner of the room some charcoal lying on a patch of burnt clay probably indicated the position of a hearth (F163) while another charcoal patch in the south of the room (7/17) could less certainly be identified in this way.

Above this general level comprising the Stage VII floor in Room 31A, was a series of small, mostly interleaving patches of clay, clay-loam, loam, charcoal or charcoal-loam. Cutting some of these patches, but sealed by others was gully F164. This was flat-bottomed, *c* 0.40 m deep with almost vertical sides and a slightly rounded southern end. Its northern end had been cut by the robber trench of the northern wall of the range, but the gully did not appear to the north of this, and was, therefore, not a drain. Its fill of mixed loams gives no clue as to function, but it is possible that F164 had once been the foundation for a dividing wall in Room 31A.

Room 32 was floored with a pebbly loam (7/31, 7/32) which contained more loam than similar floors at approximately the same absolute level within the North-west Range. Of floors at this level, that in Corridor 33 contained least loam of all. There the pebbles (5/4) had been laid against the west wall of Room 42 and consequently post-dated the construction of that Room, although not necessarily by very long.

Dating evidence for Stage VII (PVW)

The high proportion of residual material in this stage may be noted. No 190 is however probably early third century. There are a number of much later pieces in the demolition deposit (nos 230-33, 250) and these probably reflect later activity on or near the site. The majority of the pottery from the site should or could pre-date *c* AD 260.

STAGE VIII

Most evidence for Stage VIII is associated with the Bath-house, which no longer served its original function. Rooms 42 and 45 were resurfaced with similar material to that found in Yard 46. In the North-west Range the latest, very patchy layers, suggest less intensive occupation than previously (Fig 17).

The plan of Stage VIII also shows the sequence of robbing which may or may not have occurred before the alterations described here.

Bath-house

Verandah 38 was replaced by Room 52 formed by walls F188-190, one of which (F188) also redefined the eastern side of Room 51. Between wall F189 and the southern limit of excavation was a short length of wall (F191) which, with no internal surface to its east, must have returned westwards to form a room south of the Bath-house.

The new walls were built without foundations, the bottom course resting on the pre-existing stratification, apart from that of F189 which was supported by a stone and clay packing where it crossed subsidence into the pre-complex ditch F0. The walls were *c* 0.50 m wide and consisted of Bunter Sandstone slabs laid mostly as stretchers, either side of an infill of smaller pieces of the same stone mixed with loam. Generally only two courses remained, although four were recorded above the subsidence.

Stratification in Room 52 was fragmentary but several tiny patches of grit and gritty loam with pebbles, all butting against and therefore later than walls F188, F189, are interpreted as building debris used as make-ups for equally patchy floor finishes (3/7, 3/8, 3/21, 3/22). Cut through these patches were two small pits, F194 and F196, filled with brown loam.

The new room implies the retention of at least the eastern parts of the old Bath-house. In the north-east corner of Room 51, pebble-loam floor 2/19 abutted and was later than the west wall of Room 52 (F188) and dipped into the east end of the main Bath-house drain, which was therefore uncapped and out of use. Floor 2/19, so also Room 52, must have been built soon after the Bath-house lost its original function.

Other internal detail of the Bath-house at this Stage is lacking, most of the area having been removed by medieval robbing.

Rooms 42 and 45

Overlaying the Stage VII surfaces of Rooms 42 and 45 were substantial floors of compacted gravel and loam (8/1, 8/10, 8/25), very similar to the major surface of Yard 46 (1/15), laid after Room 52 was finished.

In Room 42, surface 8/10 was cut by a sub circular

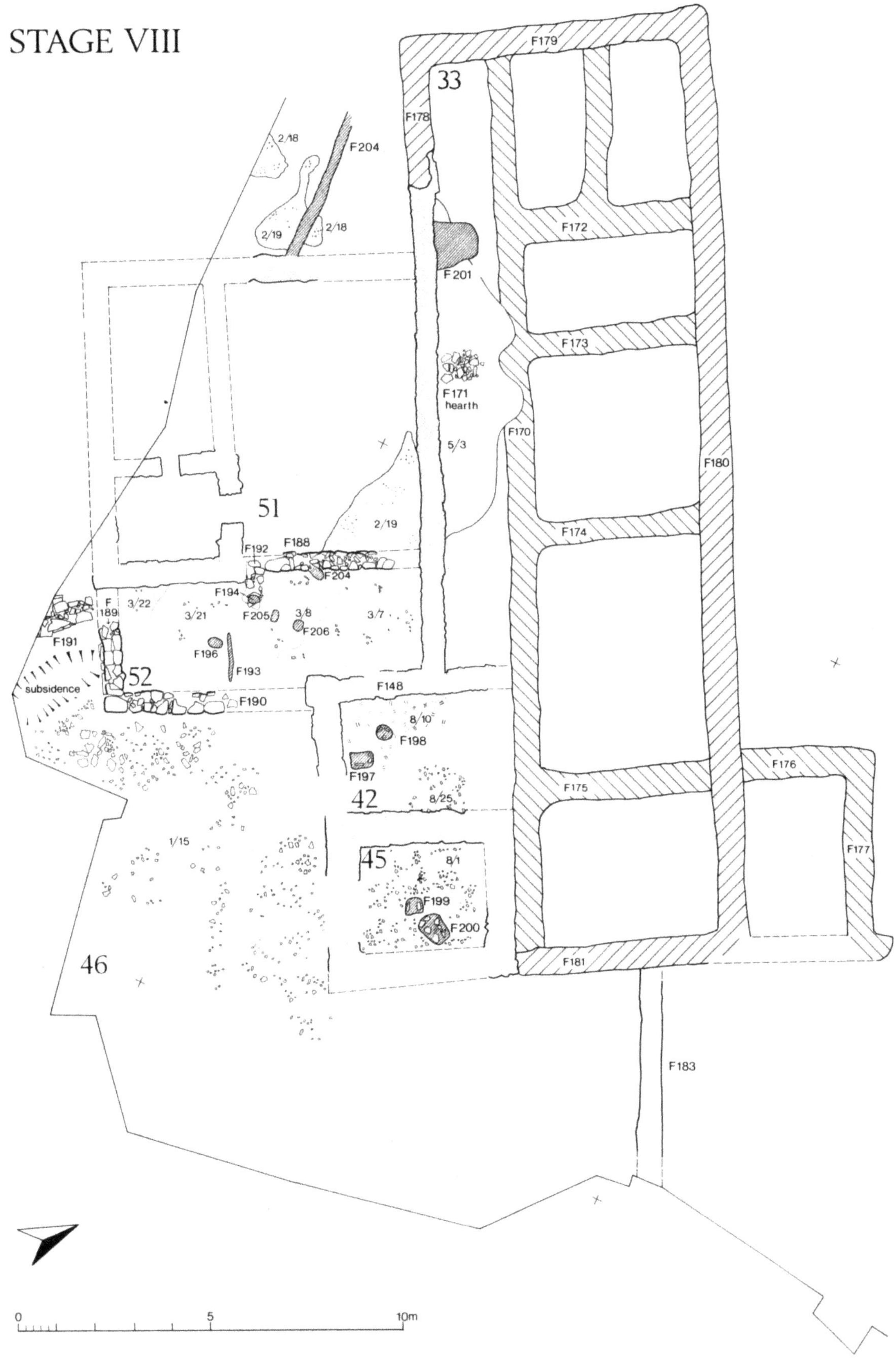

Figure 17: Plan of Stage VIII, scale 1:150

post-pit F198, with packing stones still *in situ*. A similarly built, sub-rectangular post-pit just to the south-east (F197), cut Stage VII levels only. Two more pits (F199, F200) of the same type cut the gravel floor (8/1) in Room 45. The stratification in both rooms was cut at the edges by robber trenches (possibly medieval) and sealed horizontally by ploughsoil.

External Areas

Yard 46 was completely resurfaced with a substantial gravel and loam deposit (1/15), which was later than the west wall of Room 52. Subtle changes in colouration and loam concentration suggest that the layer was formed in several tips which were then presumably levelled before more gravel was laid on top. Away from the walls the ground was more compacted. The northern edges of 1/15 were cut by robber trenches of the southern walls of Rooms 42 and 45; elsewhere the surfaces was directly sealed by ploughsoil, and penetrated occasionally by modern farmyard pits.

The North-west Range

No change was detected in Room 32 but in Corridor 33 were two hearths (F171, F201). The more westerly, (F201) was made of purple and yellow sandstone slabs lying in a charcoal-loam matrix capped by charcoal. The stones in the hearth and the surviving stones of the wall immediately behind it had been burnt red. Hearth F171 was similarly constructed but may have functioned differently as its stones and those in the wall were blackened, not reddened. The charcoal capping F171 (5/3) spreading to its north was recorded as sealing the southern edge of robber trench F170, over two small areas of *c* 0.20 x 0.50 m. In the robber trenches however, it was often difficult to distinguish between robber trench backfill and slivers of construction trench backfill adhering to the side of the original construction trench on almost the same vertical plane as that of the robber trench. If the recorded relationship really does refer to the robber trench fill being overlapped by F171, and not merely the construction trench backfill, the implication is that F171 was built after the demolition of the complex, and while its ruins were sufficiently exposed for the adjacent wall to become charred. The position of the hearths implies the disuse of Corridor 33 as a route of access, so if they were in use in a standing building, the North-west Range must have been close to evacuation, if not demolition. Hearth F201 was sealed by a layer of destruction debris which in turn sealed the western part of the site, marking the final abandonment of the complex.

North Range

There is no evidence to show whether or not this range was still in use in Stage VIII. Without positive evidence to the contrary its demolition is assumed to have taken place at the same time as the rest of the complex.

Alterations during Stage VIII

Room 52

The latest structural alteration on the site was the division of Room 52 into two roughly equal parts. Lying above the pit/posthole F194 which cut a surface (3/8, 3/21) laid against the Stage VIII wall (F188) was a short length of east-west walling (F192). It was made of slabs of micritic Bunter Sandstone like those used in the Stage VIII walls. The Manchester University excavations of June 1976 had removed a little of the stratification from this room and a shallow trench (F202) was noted, which extended the alignment of wall F192 eastwards. The trench lay immediately below topsoil and cut the latest Roman stratification in the room; it is therefore interpreted as the robber trench for part of wall F192. A shallow slot with the same stratigraphic relationships, just to the south and east of robber trench F202, may also have been associated with the division of the room.

A patchy clay-loam had been laid against both sides of wall F192, forming the new floor of the room. Several large pieces of tile and flat stones pressed into this in the northern part of the room could have been part of this floor surface or later repair. Also in the northern part of the room were three small pits (F204-6) which cut the floor. In the south-west corner of the room was a heap of wall plaster (most of it white), nails, tile-fragments and charcoal; all mixed with loam. The derivation of these materials is uncertain and they may have been brought into Room 52 simply to be used as make-up for a floor still unmade at the time the complex was abandoned. Stratification in the northern part of Room 52 was cut by the robber trench of wall F148, which contained a sherd of medieval pottery; the room as a whole was sealed by ploughsoil or topsoil.

Dating evidence for Stage VIII (PVW)

The material from Stage VIII may be grouped with that from the VII-VIII accumulation in the ditches to the west of the complex. No 260 is probably mid third century or later. There is nothing to suggest definite Stage VIII activity after the early to mid third century.

DEMOLITION OF THE COMPLEX

After Stage VIII, much, if not all of the complex was demolished. Most of the North-west Range appears to have been robbed not long after the last occupation of the complex and the North Range may also have been destroyed at this time (Fig 17). Evidence presented below (p 39) demonstrates that there was a phase of medieval robbing in the Bath-house.

Bath-house and Rooms 42 and 45

To the west of the complex a layer of charcoal debris covered the backfilled ditches (p 40) and partially robbed main Bath-house drain, but it did not cover any part of the Bath-house. The charcoal was cut by a gully which was in turn covered by another layer of charcoal debris sealing parts of the Bath-house as well as the western half of the North-west Range. This debris was later cut by the medieval robber pit of the Bath-house.

As the rest of the Bath-house and Rooms 42 and 45 were only sealed by ploughsoil or topsoil, it cannot be proven stratigraphically that these structures were demolished at the same time as the North-west Range.

North-west and North Ranges

The robber trench fills were of similar brown loams with occasional concentrations of stone, charcoal and the like. Clear distinctions could be made between the various tip lines of the backfill, however, and a simple sequence was established. This showed that the cross-walls and north corridor wall had been dismantled first (robber trenches F172-177) followed by the side and end walls (robber trenches F178-181). Although some pieces of dressed stone remained at footing level, most walls had been robbed down to their foundations of igneous boulders. In some places even these had been removed.

In the area of the North Range modern ploughing had damaged the later levels. Only one robber trench (F183) survived and here stonework had been completely removed in contrast to those of walls F91, F92, F99 of which four or five courses remained. Robber trench F183 was backfilled with the same sort of material as filled robber trenches in the North-west Range although the stratigraphic relationship between them was indeterminate.

LATER ROMAN ACTIVITY

Several sherds of the later third and fourth centuries were recovered from the topsoil and suggest a source other than the site itself. Two sherds of this date were, however, retrieved from two pits, neither of which was properly defined during excavation. The presence of late-Roman sherds in such contexts is entirely inconsistent with the rest of the dated sequence and it is likely that they mark the position of intrusive features.

MEDIEVAL ACTIVITY

Cutting the charcoal debris sealing the eastern parts of the Bath-house as well as the Stage III floor (3/19), and sealed only by topsoil was a large pit. This is interpreted as the robbing of parts of the Bath-house. The pit was concentrated on former Room 50 (the *frigidarium)* and from its depth seems to have been dug to remove the *pilae* tiles as well as the walls and their foundations of which none remained save next to the plunge bath (F130) (p 30).

Two sherds of medieval and 'late-medieval' pottery were found low down in the fill of this robber pit. These have been reported on in full with the rest of the medieval pottery (O'Leary and Davey 1977). The two sherds found in the pit were a sherd of unknown source dated 1250-1350; a sherd of form and fabric 'similar to many found in Ewloe, Clwyd' and dated to between 1400 and 1500. Both these sherds were well sealed and must be taken as evidence for a phase of medieval robbing. This does not mean that the Bath-house may not have been dismantled to ground level in the Roman period, but for robbing to have taken place in the medieval period, the position of the Bath-house must have been clear.

THE AREA WEST OF THE COMPLEX

To the west of the complex lay a track and a series of ditches. Enough stratification remained to establish a simple sequence which is presented in the order of its apparent relationship to that within the complex itself (Fig 18).

After removal of ploughsoil all visible stratification was planned and parts of the ditches investigated in three *sondages* (A-C). Additionally, parts of some extant layers were removed to test the apparent absence of features beneath (eg parts of track 4/14).

Stages I-III

Stratification pre-dating ditch F500

Two of the earliest layers in this part of the site (F/13, 4/17) have already been described with the pre-complex ditch F0 (p 8). In *Sondage* A, lying on, and partially pressed into the natural subsoil was a patch of hard-packed pebbles which may have formed part of a track (4/14).

Ditch F500

This was a flat-bottomed, U-shaped ditch, fully excavated in *Sondage* A only. It was defined in plan running parallel with the western wall of the complex, before passing beyond the western limit of excavation. Its primary filling contained eight fragments of brick and tile, materials that did not appear on the site until Stage IV while F500 may have cut *de novo* or completely recut along the same line.

Stratification later than ditch F500 but earlier than Stage IV

The primary fill of ditch F500 was overlain on the eastern side by pebbles 4/14. They rested on a thick and slightly cambered clay probably redeposited from a cutting of the ditch F500. The track had been cut on the south by Stage VI drain F506 and on the east by the Stage IV North-west Range. On the north it ran beyond the limit of excavation, although here, in contrast to its dense and well-compacted surface near the buildings its comprised only a few scattered pebbles.

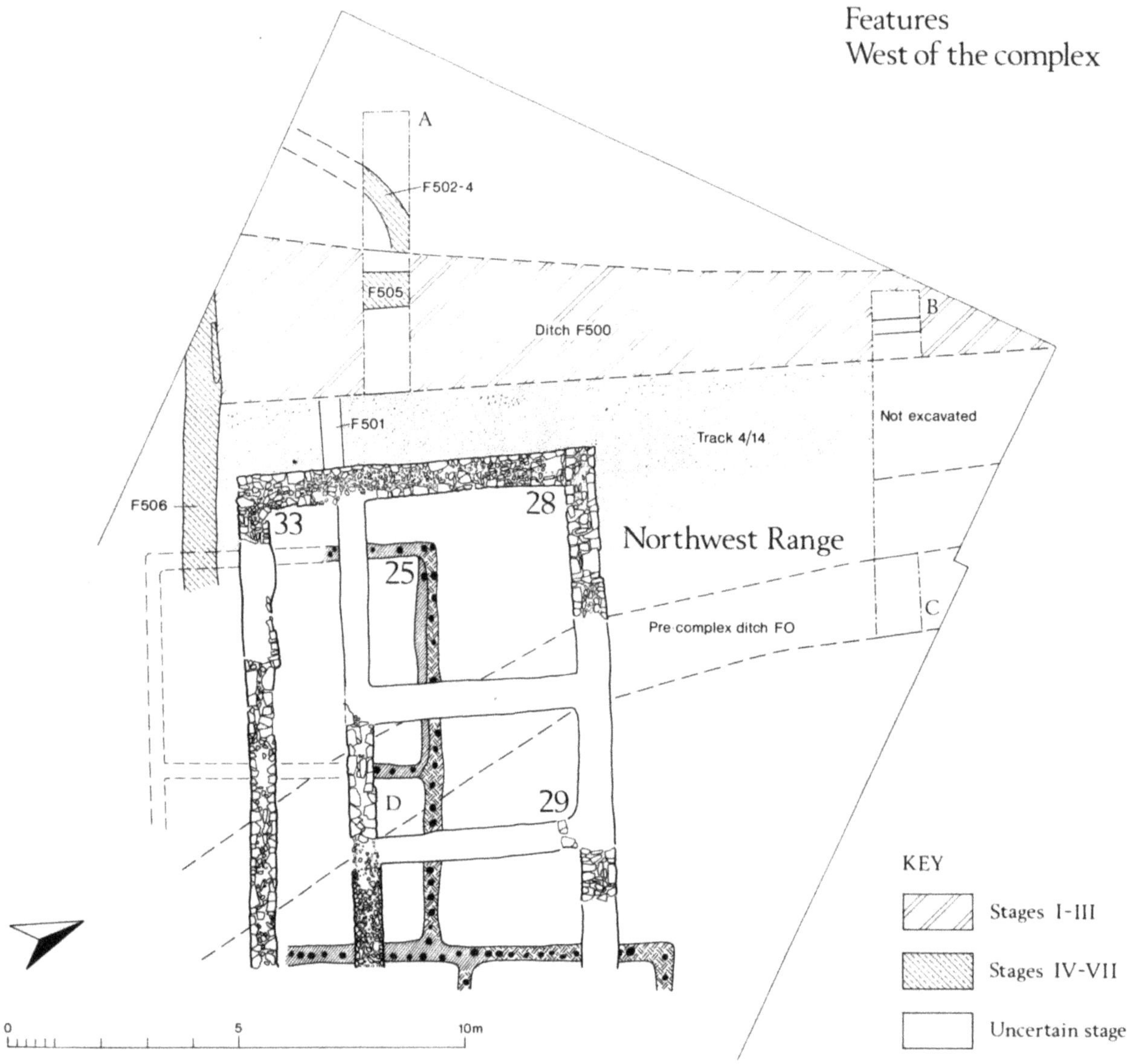

Figure 18: Plan of features west of the complex, scale 1:150

Stages IV-VII

Ditch F505 and subsequent gullies F502, F503, F504

Ditch F500 was cut by ditch F505 the square profile of which closely resembled that of the Bath-house drain F506. One flagstone lining drain F506, which had survived the later robbing activity, was almost identical with another found disturbed in the upper fills of ditch F505 in *Sondage* A. The two features were presumably part of the same Bath-house drain, stone-lined at least as far as *Sondage* A, although in *Sondage* B the ditch as a whole, and the bottom of it in particular, had narrowed. Similar primary silts had accumulated in both these drains.

In *Sondage* A, a subsidiary gully, F502, fed into ditch F505 curving from the extreme south-west corner of the site. Butting the backfilled F502 was a shallower gully of similar profile F503, which, when itself backfilled, was cut to form a third, similarly profiled gully (F504). The fills of these gullies were very similar to, and may have been continuous with layers of demolition debris which generally sealed this part of the site (see below Stage VII).

Stage VII and later

All but one of the topmost layers to the west of the complex were characterised by their high charcoal content and profuse flecks of brick or tile, the single exception was a soft, gritty loam that filled the Bath-house drain (F506) after its lining had been partially robbed.

The earliest of the charcoal debris layers filled ditch

F505 above this primary silt. This was overlain by a layer containing much Hawarden Sandstone - common to Stage IV, which levelled off the area.

Ditch F505 was capped by a sequence of which sealed all of the ditches and gullies, the track (4/14), and robber trenches in the western parts of the complex (F172, F173, F176 part, F180 part, F178) as well as parts of the re-used Bath-house.

THE AREA NORTH OF THE COMPLEX

The sequence to the north of the complex was established through a number of intercutting ditches, gullies and pits. There was, however, very little horizontal stratification, so that detail is lacking and the sequence is presented in the order of its apparent relationships to that within the complex itself (Fig 19).

All of the features described here were revealed in plan and most were excavated. Together these features represent a small yard to the north, and to the north-west of the complex.

Stages I-III

Ditch F300 and earlier features

The L-shaped plan of ditch F300 cut into natural subsoil, lay on the same alignments as the complex. Between the point where it was cut by later features just north of Room 2 in the timber North Range and the point where it ran out of the excavated area at the NW, the bottom of the ditch sloped downwards by about 0.25 m. The primary fill was a silty loam. A second ditch (F301) sloped downwards *c* 0.27 m from its point of origin in Room 2, where it cut backfilled pit F4,to its junction with ditch F300. As it did not appear on the northern side of ditch F300, where there were no later intrusions to interrupt its continuation, of the two ditches presumably worked together with F301 feeding into F300.

Ditch F301 was filled with clays and mixed charcoal and clay lying beneath a clean black charcoal deposit capped by hard orange clay partially burnt red, which levelled-off the ditch. The northern end of these fills (below the black charcoal) was overlain by the second deposit found in ditch F300. As the interface between the fills of the ditches formed a sharply defined edge, it is implied that ditch F300 was recut after ditch F301 had been filled up, but before the charcoal was deposited and it had been levelled off with clay. Ditch F300 was subsequently filled up with orange clay, mottled with pinker lumps apparently burnt; this closely resembled both the top fill of ditch F301 and a layer in the middle of the yard (12/1) cut by later features.

Clay of this type occurred exclusively in layers which represented the demolition at the end of Stages III and VI. In each of these the levelled clay walls lay on top of extensive charcoal deposits and were often mottled pink or red presumably as a result of exposure to heat. In the case of ditch F300 the fill was itself cut by a Stage IV ditch (F304), so that the levelling of both ditches F300 and F301 can be reasonably assigned to the demolition work at the end of Stage IV.

Also cut by the Stage IV-V ditch (F304) and cutting natural subsoil was a small rectangular pit (F302) of which only small amounts remained on either side of the ditch.

Stage IV-V

Ditch F304 and construction of stone-lined pit F303

Where it cut ditch F300, ditch F304 appeared to have been dug as one with a deep stone-lined pit (F303), itself cutting ditch F301. The south-eastern corner of F303 was overlain by clay that had been spilled from the base of the north wall of Room 40 (Stage IV). The construction pit for the stone-lined tank cut into the natural ground and soft yellow sandstone to a depth of 1.50 m below contemporary ground level. Its sides were then lined with between six and eight stretcher courses of dressed Hawarden Sandstone of a type in use from Stage IV onwards. On the other side of the pit, where the builders had not encountered the solid foundation of sandstone bedrock that existed on the southern side, two extra courses had been used. Near the top of the east end of the lining was a rectangular socket formed by standing two stones on their sides,the function of which remains unclear.

In the pit was a dump of silty sand *c* 0.80 m deep the top of which lay just above the level of the bottom of the stone lining. A similar but shallower deposit covered this and was indented by a layer of angular pieces of sandstone set on end and pitched slightly eastwards. They were covered in turn by a layer of pebbly sand. These four deposits contrasted strongly with the silty loam immediately above which contained nails, brick and tile fragments, stone and charcoal representing disuse. Immediately above this layer on the eastern edge of pit F303 was a gravelly sand that had spilled in from the bottom of ditch F304, where it lay beneath the Stage V lead pipe F100 (p 28). All subsequent fills of ditch F304 were later than the insertion of the pipe. The interpretation of pit F303 is unclear but it may have been a soakaway or cess-pit.

Lead pipe F100

A short length of lead pipe (F100) protruded from two layers of hard pinkish clay which formed part of the base of the north wall of Room 40 built in Stage V. The splayed end of the pipe led into the south of ditch F304. Just above and immediately beyond the splayed northern end of the pipe was a matrix of gravel pebbles and loam.

Stage V-VI

Gully F305

A shallow gully (F305), falling slightly from west to east some 2 m north of Rooms 39 and 41, curved around the stone-lined pit (F303) and fed into ditch F304 near its southern end. The gully became

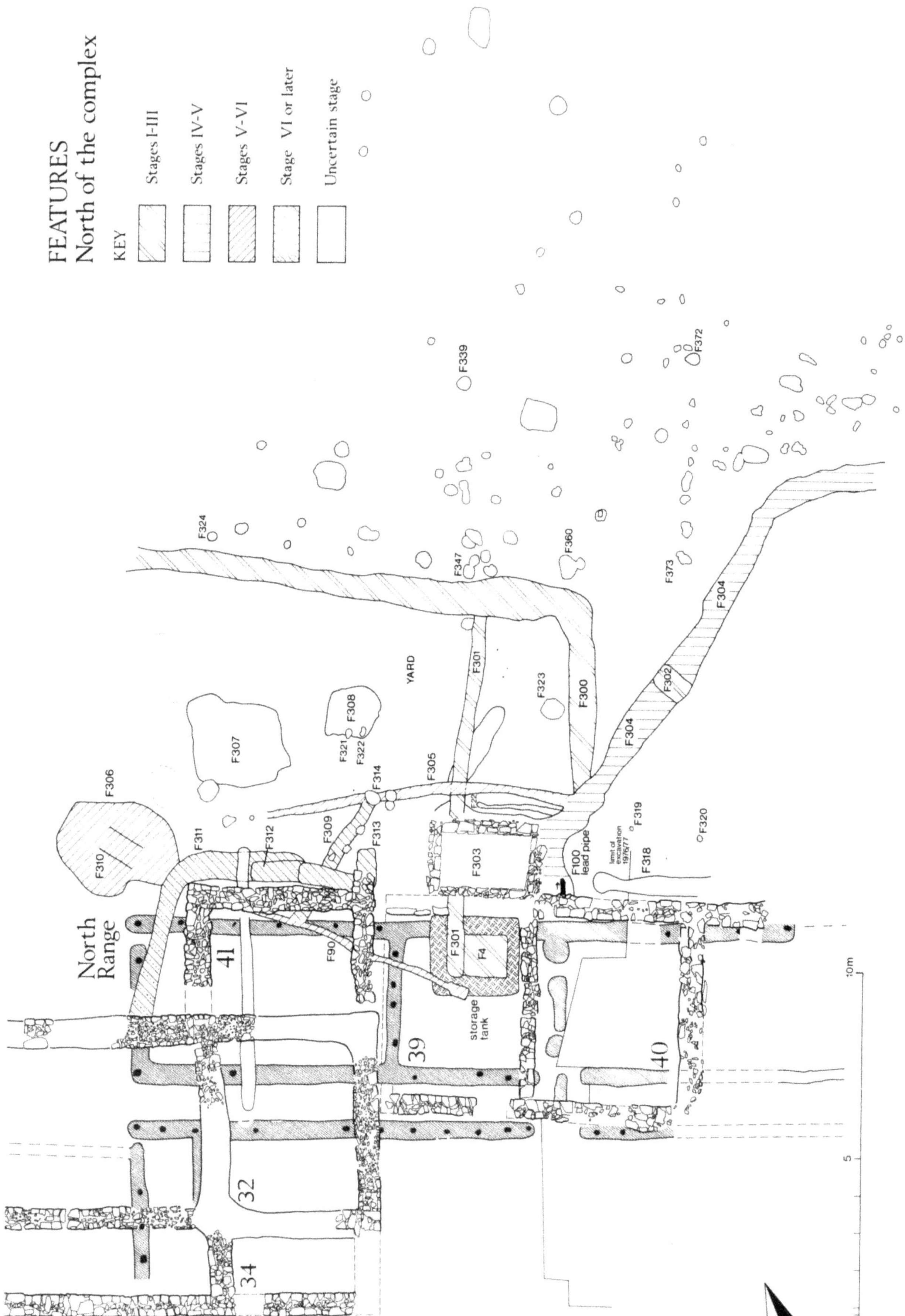
FEATURES
North of the complex
KEY
Stages I-III
Stages IV-V
Stages V-VI
Stage VI or later
Uncertain stage
North Range
YARD
storage tank
F100 lead pipe
limit of excavation 1976/77
F4
F90
F300
F301
F302
F303
F304
F305
F306
F307
F308
F309
F310
F311
F312
F313
F314
F318
F319
F320
F321
F322
F323
F324
F339
F347
F360
F372
F373
32
34
39
40
41
0
5
10m

shallower the further east it went: presumably it had been truncated by ploughing. In the eastern part, clay and gravel filled the base of the feature: the rest of the gully was choked by charcoal. This charcoal fill was cut by gully F309 ascribed to Stage VI or later.

Drain F310 and Pit F306
Towards the west of the yard was a drain (F310) which linked the end of a stone-lined drain (F90) running from Room 40 to a possible soakaway pit (F306) which was not fully excavated. As drain F90 was of one build with Room 40, drain F310 and pit F306 must have been built by Stage V at the latest. Apart from the primary fill of drain F310 the fills of both features were similar interspersed clayey loams and silts. Above these fills, over an area slightly larger than the diameter of the possible soakaway (F306) were two deposits, which may have resulted from sinkage into an incompletely filled soakaway. They were cut by gully F311, attributed to Stage VI or later.

Stage VI or later
Gullies F309, F311; Pits F312, F313, F314
Cutting the capping layer of pit F306 was gully F311. It drained from just north of Room 31A to a point just outside the north wall of Room 41, where it had a rounded end. Its line shows that it was later than Room 41, which was itself a secondary addition to the Northwest Range. Its silty primary fill might have accumulated during its use but it had then become choked by a secondary fill of pebbly loam and charcoal.

The rounded north-eastern end of gully F311 was cut by pit F312, the silt loam fill of which was cut by another pit, F313. Its fill was similar to that of F312 but had clay inclusions.

A small gully running north from Room 41 (F309) cut choked drain F90 and was itself cut by pit F313. From the relationships with drain F90, gully F309 must belong to Stage VI at the earliest. It intersects F305 at the point where each were cut by a later pit (F314).

Disuse of stone-lined pit F303; fill of ditch F304
Ditch F304 was evidently in use when fed by lead pipe F100 of Stage V. At a later date the ditch was filled with deposits linking with the uppermost fills above stone-lined pit F303. Sediment analysis of the earliest of the pit-fills suggested that it was not water deposited but formed by a natural process of accumulation, the breakdown of soils and other sediments, and the addition of much detritus. The overlying deposit was concentrated on the southern side of pit F303 where it had spread from Room 39, either because the northern wall of the room was incomplete or because there was a doorway. This deposit, of clayey loam with charcoal and tile flecks resembled the Stage VII floor in Room 39 (9/14) and may have been surplus to the reflooring process.

Above this, two charcoal-charged, clayey-loam deposits contained several lenses of clayey loam and charcoal containing lead oxide, galena, lead and bone fragments. Although these lenses contained much burnt debris, sediment analysis showed that quartz grains themselves had not been burnt suggesting the lenses were domestic rubbish tips rather than the burnt building debris. Indeed all three layers imply dumping and/or accumulation of rubbish in the hollow above the now-defunct pit (F303).

Although none of the deposits just described physically joined any of the fills of ditch F304, correlation is possible. One of the ditch fills contained galena etc resembling levels in pit F303. A charcoal-flecked loam with glass and pot fragments, and other finds also occurred in both features (F303 and F304), although in ditch F304 it comprised a major proportion of the fill. Finally, however, the two were filled to ground level by a continuous clayey loam, containing brick and tile fragments, pot, daub, lead oxide, tiny greenish glass globules and a little charcoal. Thus joined, F303 and F304 were cut by three small pits.

FEATURES OF UNCERTAIN STAGE
Isolated deposits to the east of F304
Between ditch F304 and the eastern edge of the excavated area was a sequence of deposits not directly related to features described above. The two earliest layers both consisted of charcoal, flecked orange clay, perhaps derived from the demolition of clay walls from adjacent Rooms 1 and 2 demolished prior to Stage V. The overlying patches of gravel and lead oxide definitely post-dated the construction of the Stage V North Range, because some of the lead oxide and gravel butted against and recessed into the vertical joints of the wall. Through these patches was cut a gully (F318), running parallel with the north wall of Room 40, which may have served as a storm-water gully before being filled by charcoal-charged loam and tile and charcoal-flecked clay. A similar layer then covered the area immediately north of Room 40 before being cut by two small stakeholes F319, F320. The tops of these deposits and the top of ditch F304 were truncated by medieval and later ploughing.

Isolated debris south of ditch F300.
South of ditch F300 and north of the complex were several pits. Pits F307 and F308 both cut the pinkish orange clay attributed to the destruction of clay walls at the end of Stage III. Pit F307 was not excavated, but the loams fill of pit F308 which contained bones and

Figure19: Plan of features north of the complex, scale 1:150

lead oxide, was cut by two possible stakeholes F321 and F322, both filled with brown loam.

To the east of these features, cutting natural subsoil and sealed by topsoil was pit F323.

Isolated features north of ditches F300 and F304
Almost all of the features to the north of ditch F300 were found immediately below ploughsoil and, without intervening stratification, cut natural subsoil. The exceptions lay near the east of the excavated area where they were covered by an extensive layer of blackish brown silty loam containing pot and some brick and tile fragments. Sediment analysis led Dr Shackly to conclude that this had been formed before its deposition here by the same demolition processes that created similar deposits west of the complex.

Fig 19 shows the hundred odd pits and possible post-pits in this area. Analysis of their dimensions and fills failed to produce any obvious patterns. None showed a post-impression. Even along the three most obvious alignments running east-west between F324 and F360; north-south between F347 and F339; and north-south between F373 and F372, the pits showed neither uniformity of dimensions nor similarity of fills. The east-west alignment between F324 and F360 may well have formed a fence immediately north of drainage/boundary ditch F300. Ditch F300 has been interpreted as a boundary and the two may have functioned together. Pits along this alignment had been recut at least once. The two north-south alignments were more difficult to interpret, even in such a general way. It may be noted, however, that their alignments were approximately parallel to one another.

Moving north the pits became fewer and virtually none occurred in the area of possible cobbling (described below pp 44-45), the southern edge of which was very roughly parallel with the complex, reinforcing the suggestion that the stone layer which made up the 'cobbling' was not a geological formation.

THE AREA IN THE NORTH OF THE SITE

The first phase in the proposed redevelopment of the Pentre Farm site was the insertion of a T-shaped cul-de-sac off the Chester Road, so that building could take place in the surrounding plots. In the north of the site a large endangered area was defined by the 'cross' of the cul-de-sac. Professor Jones' excavations of June 1976 (Fig 4) had tested this area, uncovering several postholes, included amongst those described in the preceding section.

The possibilty that these may have belonged to an arrangement of timber structures only comprehensible in the wider view led to the mechanical stipping of ploughsoil and modern debris from the whole of the 'cross' of the cul-de-sac (about 750 sq m; henceforth the north of the site). Depth varied between 0.35 m in the west and 0.60 m in the east where the overburden had been increased by the debris from the recently demolished farm buildings. The area thus uncovered was stripped, in three consecutive spits to the top of the subsoil. The only features detected were a twentieth century ditch which had lain alongside the pig-sties of the old farm in the extreme east of the excavated area, and in the extreme west, a few small postholes which lay in the vicinity of the westernmost *sondage* investigated in the Manchester University excavations. Uncertainties about the interpretation of the postholes prompted more detailed area excavation, 15 m x 7 m, in the extreme north-west area of the 'T', the size being determined by the limits of a plastic shelter used to protect the excavation from the severe winter conditions.

The Ditches
The earliest ditch F12 was filled with a virtually uncontaminated natural loam, presumably derived from the weathering of the sides, before being recut along its length by a deep and narrower U-shaped ditch (F2). In turn, before any fill had been deposited in it recut ditch F2 was cut by east-west ditch F4. The two must have functioned together because water running from south to north in ditch F2 could not have escaped along the silted continuation of ditch F12. The base of ditch F4 was, however, at a similar level as that of ditch F2 and its slope would have carried waste away to the east.

The homogenous fill of F2 interleaved with those of a short length of curving U-shaped gully (F3), which fell gently from south to north. Gully F1/F14 linked gullies F2 and F4. The upper parts of the fill at the northern end of the ditch (F2) were cut by ditch F16, which lay to the north of ditch F2 and parallel with it. It narrowed but increased in depth as it neared ditch F4, which must have remained in use with ditch F16.

The scatter of igneous boulders and other remains
Within the area excavated to the south of ditch F4 and to the east of ditch F2/F12, was a layer of glacial igneous boulders, usually no more than one deep and often packed closely together. They lay above and partially within the natural subsoil and beneath the ploughsoil, although in the west they were also covered by a debris layer. Occasional boulders did run to the west and north of the ditch (F2/F12 and F4), but the marked increase in density to the south and west of them suggested that the boulders may have been purposely laid inside an area defined by the ditches, although nowhere was there a clear stratigraphic relationship between the two. A similar change in density

Figure 20: Plan of ditches and cobbling north of the complex, scale 1:150

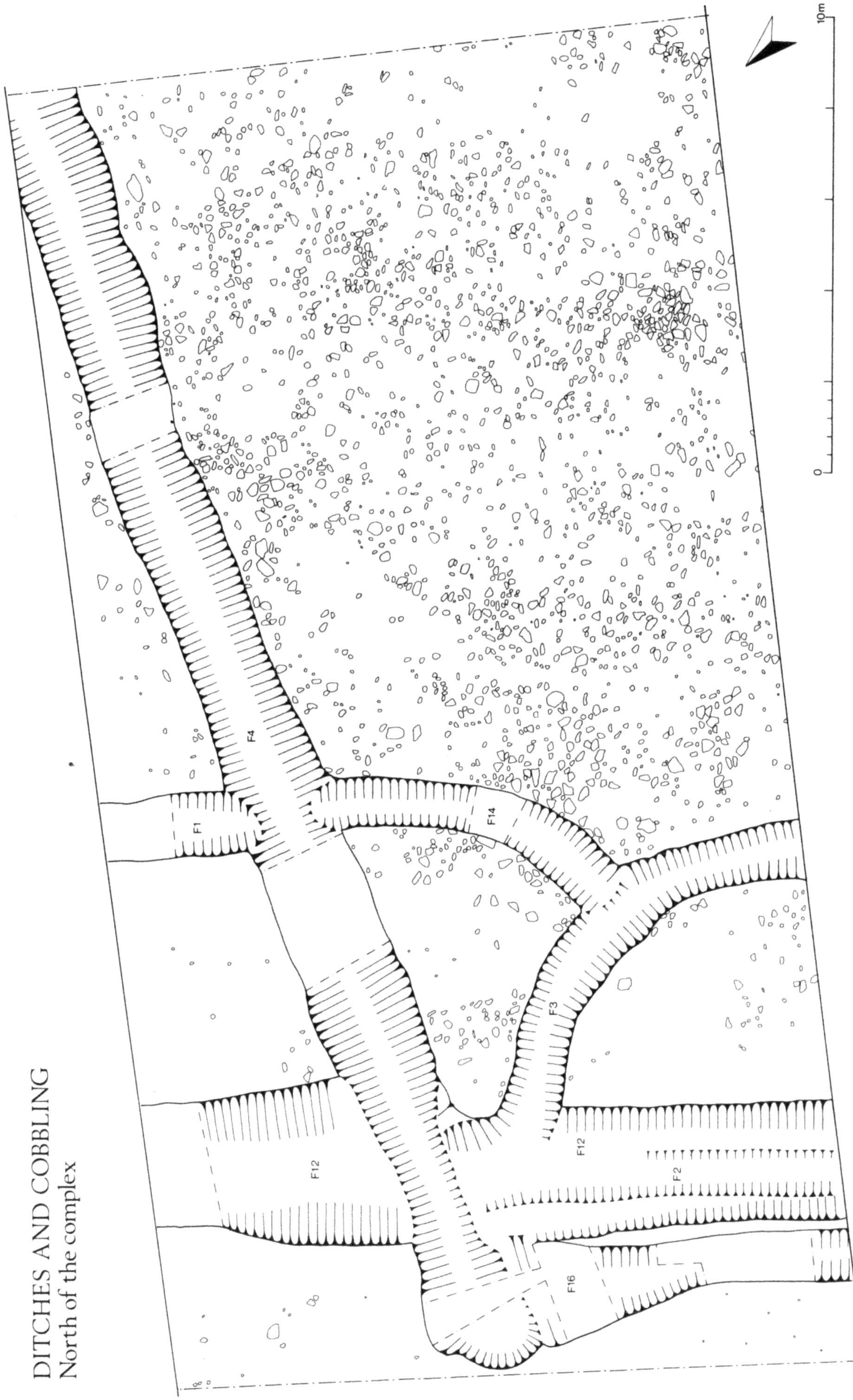

DITCHES AND COBBLING
North of the complex

occurred just north of the complex at the point where the boulders merged with a mass of small pits and/or postholes (p 44).

Sealing ditch F2 was a localised charcoal flecked loam deposit cut by three of the stakeholes discovered by Manchester University in July 1976, and just north of the loam deposit were three more stakeholes, one cut into the natural subsoil and two into the fill of ditch F2.

Phasing and dating of the ditches

On the basis of artefacts from the ditch fills, the features in the north of the site can be related to the interpretive framework of stages worked out for the complex.

The fills of ditch F4 contained pottery including a sherd of a flanged bowl in Black-burnished Ware (AD 120-60) from the primary silt. Also in an upper fill of ditch F4 were fragments of twenty yellow-brown Hawarden Sandstone, a type of stone not introduced to the complex until Stage IV.

In the fill of ditch F2 were several fragments of tile, a material which does not seem to have been introduced to the complex until Stage IV or just before. There was also a sherd of Central Gaulish Samian ware, Form 37 (Hadrianic-Antonine).

To summarise, whilst F12 might have pre-dated the construction of the complex, ditches F2 and F4 seem to have been broadly contemporary with Stages I-III. The subsequent intercutting and filling of ditches continued until at least Stage IV and possibly later.

INHUMATIONS

By Guy Granger (1981/2)

Two inhumations were located north of the complex, during the construction of a garage (Fig 4). Both were aligned north south, with heads to the south. Machine clearance of the surrounding area revealed no other burials. The proximity of the second to the first implies knowledge of the position of the first burial when the second took place. The fact that Hawarden Sandstone was used in the lining of one of the graves (first appearing on the site in Stage IV) and tiles used in the other (first appearing on site again in Stage IV) suggests that the burials date to Stage IV or later. A date of *c* AD 150 or later is therefore suggested for the burials.

Burial 1

In the area of the burials the natural glacial deposits of the Dee estuary lay beneath a mid- to red-brown soil formation. This had formed to a depth of *c* 0.10 m by the time it was cut by Burial 1 which measured at least 1.8 x 0.45 m. The eastern side of the cut had been removed by the later grave (Burial 2) but the western side was intact and sloped at an angle of 55° through the unstable pebbly subsoil.

The base of the grave was covered by a line of four *tegulae*, outer faces uppermost. A fifth *tegula*, outer face inward, stood on edge to the north of these. Although the eastern side of the grave had been disturbed the western side and southern and of the grave certainly contained no tiles on edge.

Above the western half of the northernmost tile was a deposit of pebbly subsoil up to 0.10 m thick. The left leg of a skeleton rested on this and was thus raised above the level of the other skeletal remains which were badly decomposed and partially crushed. The arms of the skeleton seem to have lain at the side. The skeleton was buried beneath redeposited pebbly subsoil (0.30-0.40 m thick). This in turn was capped by another line of four *tegulae*, outer faces uppermost. Unlike the tiles on the base of the grave, these were not lying flat but had broken in several places where they had subsided into a depression in the top of the underlying subsoil backfill (*c* 0.25 m deep in the centre). The uppermost 0.30 m of the grave was filled by a mixture of humus and redeposited pebbly subsoil.

Skeleton remains, Burial 1

The condition of the burial was very poor; the skull and pelvis were fragmentary and useless for determining the sex of the skeleton, the ribs and vertebrae only survived as spongy fragments, and the long bones were fragmented with the ends severely eroded.

Dentition

R		L
- - - 5 - - - -	- 2 - - 5 6 7 8	
8 7 6 5 / 3 2 1	1 2 3 4 5 6 7 8	

/ missing ante mortem

Dental attrition suggested an age of between 25 and 35 years, though only molars on the left side of the mouth could be inspected. The molars in the right side of the mandible were covered with a moderate to severe layer of dental calculus on all surfaces including the occusals, which would suggest that the upper right molars had been lost sometime before death. Other teeth were almost free from dental calculus. Moderate to severe dental hypoplasia was present on all teeth, and a slight to moderate degree of periodontal disease was seen.

A single mental foramen was seen on the left side of the mandible, the right side not surviving. No mandibular tori were visible.

The skull had minimal brow development, and small to medium mastoid process and an external occipital protruberance, all of which indicate that the individual was either a female or a very lightly built male.

The most complete long bone was the left femur, the extant but incomplete length of which was 345 mm. Photographs indicate the total length of the femur was *c* 380 mm which in turn (using the table of Trotter and Gleser 1958) would suggest a height of 1.49 m.

It is therefore most likely that this individual was a female.

Burial 2

Parallel to the eastern side of Burial 1 was a second. It was cut from the same absolute level as the first but was half as big again (2.9 x 0.8 m). The north-west edge of this cut had exposed the north-west sides of tiles forming the base and cover of the earlier burial.

The full circumference of the large, new cut was lined with dressed blocks of Hawarden Sandstone standing on edge. On the western side two blocks of a second course remained.

On the base of the grave were the remains of a skeleton which, like the earlier one, was decomposed and partially crushed. Its head lay at the southern end and there was a gap of *c* 1 m between its feet and the northern end of the grave adjacent to the disturbed part of the earlier grave (Burial 1). Its arms lay at the side.

Above the skeleton were four pieces of re-used lead sheet, one of which was marked with stars and crosses, pierced in several places with iron nails (pl XIII), perhaps indicating that the body had been placed in a wooden coffin.

The lead was bent inwards: northern and southern edges being *c* 0.25 m higher than the central parts.

The lead was covered by two courses by Hawarden Sandstone blocks pitched on edge which filled and slightly protruded from the top of the grave. Humus and a few pieces of broken roof tile were mixed with them. That the stones projected above the grave cut indicates that they may have been intended to mark the position of the grave.

Skeletal remains, Burial 2

This burial was in roughly the same condition as Burial 1 though the bones were even more fragmented.

Dentition

```
  - - - - - - - 1 | 1 - - - - - - -
R ----------------+---------------- L
  8 7 6 5 4 / / 1 | 1 2 3 4 5 6 7 /
  R   C C           C
```

/ missing ante mortem
C dental caries
R root only

Dental attrition suggested an age of between 30 and 40 years. A slight degree of dental calculus, moderate dental hypoplasia, with marked discolouration and banding of the teeth, and moderate to severe periodontal disease were all present. The right mandibular second molar was misplaced outwards. Single mental foramina were seen on both the left and right side of the mandible and no mandibular tori were present.

No osteological evidence for either the stature or the sex of the individual survived but photographs of the burial *in situ* show a fairly small individual, so that very tentatively it may be said to be female.

GENERAL DISCUSSION

The internal evidence for the function of the site is broadly limited to the following categories: method of construction with its apparently military connotations in the early stages; plan-form and size, again with possible military or official connotations; the presence of stamped military tiles; the structural continuity demonstrated by the stratigraphy and, lastly, a date-range of *c* AD 120-240. This internal information is not, in itself, sufficient to generate a thorough discussion of the complex and attempts at further explanation must draw on external evidence and general argument, a process which necessarily remains open-ended.

Method of Construction

The method of construction used in Stages I-III has been described best by Richmond (1961) and can be paralleled in a general way in many military contexts: Hod Hill (Richmond 1968), Fishbourne (Cunliffe 1971), Richborough (Cunliffe 1968). Nearby, at Chester, on the Old Market Hall site were slots cut into the natural sands and sandstone back filled with spoil to hold the uprights firm (Petch 1970). Enough of the walling remained to show that the spaces between the uprights had been filled by clay packed onto wattles attached to the timber uprights. Such walling has been found on many military sites: Inchtuthil (Frere and St. Joseph 1974), Gloucester (Hurst 1972), Pen Llystyn (Hogg *et al* 1968).

Plan form and size

The appearance of this method of construction primarily on military sites may be the result of a bias in fieldwork towards forts and stone structures, and away from *vici* and timber structures. Where timber structures were discovered in the civil settlement at Wilderspool (Hinchcliffe 1975), however, they were individually much smaller than the Pentre Farm complex. The same is true of the buildings in the *vici* at Chester (Salway 1958) and Corbridge (Daniels 1959). Similarities in size with Pentre Farm only occur with the so-called official residences or *mansiones* (discussed below), although a building interpreted as a *mansio* at Greta Bridge (Wilson 1974) was much smaller. Buildings supposedly of this type at Chester, Corbridge, Brecon Gaer (Wheeler 1926), and a later example at Vindolanda (Birley 1977) show little standardisation and better comparisons of both size and plan between buildings having the same method of construction as Pentre Farm can be found in commandants' houses within forts. In Stages I-III the Pentre Farm complex measured at least 600 sq m, an area similar to those of commandants' houses at Hod Hill I (Brailsford 1962), Fendoch (Collingwood and Richmond 1969) and Castell Collen I (Nash-Williams 1969).

Such courtyard buildings have also appeared as stone-built town houses, like those in insula VII at Caerwent (Nash-Williams 1969) and insula XIV at Silchester (Boon 1974), but at Pentre Farm the method of construction together with the type of plan and size of building seem to imply some military or official involvement during Stages I-III. Even in Stage IV, when a considerable change to the complex is marked by the addition of the 'villa-like' North-west Range any attempt at a purely civilian interpretation from this stage onwards must be balanced by the fact that some of the original timber structures still stood. The North-west Range resembles typical early stone-built villas such as those at Park Street, Lockleys, Cox Green and Frocester, but in all of these villas the rectangular structure was the sole constituent of the building. At Pentre Farm the North-west Range was merely a consolidation of a much larger whole using what may have been a very common building form of the day, itself derived directly or indirectly from the form of military barrack blocks. No military connection is, of course, implied *per se* but, close to Pentre Farm at the tilery of Legio XX at Holt a similar building, modelled on a villa of the ordinary corridor plan, was thought to be that of the officer(s) in charge (MacMullan 1965; Nash-Williams 1969). Whatever the similarities between the North-west Range and early stone villas, the *raison d'etre* for Pentre Farm is bound to be different from buildings normally assumed to have been erected as a result of agricultural prosperity for, on present evidence, Romanised native farming did not develop on the poor farmland of North-East Wales. A different stimulus must explain the existence of the Pentre Farm complex.

The Bath-house

If, by showing far fewer military connotations than Stages I-III, the North-west Range of Stage IV might imply a weakening military connection, the evidence from the Stage VI bath-house strengthens it again. The internal arrangement of the bath-house (p 29-30) permits the order of circulation to which the word *Reihentyp* refers. The slight deviations from the standard *Reihentyp* plan at Pentre Farm can be paralleled at Red House, Corbridge (Daniels 1959) where the *frigidarium* is positioned to one side of the building instead of at the end of the row. Large military bath-houses of the same type have been found at Castell Collen (Nash-Williams 1969) where the rooms were much longer, but those in the bath-house at Stockstadt (von Petrikovits 1975) were of similar size.

A typology of *Reihentyp* bath-houses derived from military sites shows that they developed from the basic three-room unit of *praefurnium, caldarium* and *tepidarium*, with the *frigidarium, apodyterium* and *laconicum* being additions to one end. The Pentre Farm building, dated by pottery to the Antonine period, can be related to this typology. It is late enough to have the *frigidarium* incorporated into the original layout and the possible *laconicum* was not of the circular variety which were Trajanic at the latest. It does not have an *apodyterium* but *apodyteria* do not appear in known bath-houses of auxiliary units until the Severan period (Daniels 1959). The bath-house at Pentre Farm thus fits the military typology of the Antonine period, which is consistent with its archaeological dating.

Closest of all to the Pentre Farm bath-house in type and size are the bath-houses at Tremadoc and Prestatyn where Daniels (1970) explained that only their smallness made them unusual. At Prestatyn, as at Pentre Farm, the military connotations of the *Reihentyp* plan-form were endorsed by the presence of Legionary stamped tiles (Blockley forthcoming).

Stamped military tiles

In Stage VI several tiles stamped LEGIO XX VV were used in the hypocaust *pilae* of the *caldarium*, and the presence of legionary tiles has often been taken to mark official status. Tiles in a building on the cliffs above Folkstone led to the suggestion '...that this building with its superb view across the channel belonged to the Fleet Commander...' (Cunliffe 1968). At Prestatyn 'neither the apparent poverty of the remains nor the smallness of the buildings ...precludes [an official connection]...on the contrary the stamped tiles surely strengthen arguments for links with either the military or the procuratorial department' (Daniels 1970).

The presence of stamped military tiles is not, however, evidence to be used uncritically. It has been suggested, for example, that some small second century buildings in Slovakia at Mušor, Oberleis, Stupava and Milanovce contradict the implication of the tiles of Legio XV Apollinaris found within them: 'the tiles coming as they do, from military units, mainly legions, have prompted Scholars to regard these as Roman forts, or, at least, as the houses of commanders of Roman troops...but apart from the tiles from military tile works, these buildings have nothing military about them' (Mócsy 1974). What seems to have been ignored by most writers discussing this subject is the possibility that bricks and tiles served as an article of commerce. Although MacMullen (1965) considers this possibility to be 'rather surprising' he uses it to explain the occurrence of military stamps in a number of unmistakably civilian sites. Amongst the impressive list he assembles is a villa near Burgstall in Czechoslovakia built of bricks from the Xth Gemina and another one in Switzerland of the first century using tiles drawn from the XXIst and XIth legions: MacMullan concedes that it is natural to look for a military function in a building made of legionary bricks, but suggests that the likeliest conjecture is that the legions exchanged surplus manufactured goods for food grown in the neighbourhood (MacMullan 1965). Present evidence suggests that tiles as articles of commerce are unlikely to have moved much more than 12 km from their place of manufacture, which in the case of Pentre Farm would allow them to have come from the known military tilery at Holt.

Mansiones and *Praetoria*

If the Pentre Farm complex is not conclusively military, the combination of plan-form, size, method of construction, type of bath-house and stamped military tiles suggests a military link strong enough to seek parallels for it amongst a variety of official buildings. The identification of a class of buildings as *mansiones* or *praetoria* was first tentatively suggested at Saalburg (Jacobi 1897). Later, a large building close to the bath-house at Brecon Gaer was explained as a *mansio* (Wheeler 1926), when a similar building at Newstead (Curle 1911) was also mentioned. Birley (1935) strongly supported this suggestion and indicated another large building some way south of the fort at Benwell. He stressed the correctness of explaining these buildings as official lodgings by citing an inscription from Mihilici in Bulgaria (CIL III 6123, originally quoted in Kalinka 1906). The inscription records the erection of two different types of buildings, *tabernae* and *praetoria* in AD 60-61 on the military roads in Thrace, the work being carried out by the procurator of the province on the orders of the Emperor (Salway 1958). Under the empire the word *praetorium* was used not only to describe the general's quarters in a camp, or the commanding officer's residence in an auxilliary fort, but for any house reserved for official use, such as an imperial villa, a governors' residence or the quarters provided outside the fort for officers on tours of duty (Birley 1935). This has more recently been reinforced: 'most places that were called *praetorium* were settlements on big roads in the provinces and owe their origins to houses built by the state and called *praetoria* which had to provide quarters for the emperor, governors and other officials on tours of duty' (Bogaers 1964).

It is clear that neither of the terms *mansio* or *praetorium* can be precisely defined, and to go beyond a basic recognition of the class of buildings is difficult. Little can be gained from the ground plans alone since there seems to be little consistency within the class. The plan of the *mansio* at Newstead, for example, is similar to those of such larger military houses as the *praetoria* of auxilliary forts or the tribune's houses in fortresses in which the rooms are ranged around a central colonnaded court. Elsewhere, the Benwell *mansio* was a straightforward aisled structure, while the building interpreted as such at Corbridge began with a form like that of a winged-corridor villa; at Chester, the *mansio* resembled the kind of villa which connected buildings were grouped around a central courtyard (Salway 1958). No clear trend or development in design appears, but all the *mansiones* just listed were identified by the presence of legionary tiles in buildings which, except at Corbridge, could be shown to have been part of the same structural scheme as the external bath-houses of the fort. (The Corbridge *mansio* was identified on the presence of legionary tiles alone.) Where no connection with military structures can be shown, it has been suggested that the character of these houses sets them apart from others in the *vicus*. 'Scale of construction complexity of fittings and decoration suggest the official' (Salway 1958).

Certain possible interpretations within this class of building can be considered as explanations for the Pentre Farm complex. Strictly speaking, a *mansio* was a lodging for the officials of the *cursus publicus* (Boon 1974) although recent work has pointed out that there is no sound evidence, as far as the first to the third century AD is concerned, for the assumption that a rest-house beside the route was designated a *mansio*. Other terms were employed in order to refer to such a building such as *taberna, praetorium* and *stabulum* (Bender 1975). Nevertheless, the imperial postal services established under Augustus must have used buildings of this class along its routes. Pentre Farm lay on the road across North Wales, which must have existed from the time Segontium was first established. The connection between Chester and Caernarvon is mentioned as *Iter XI* in the Antonine Itinerary, and the aerial photograph taken by Professor G D B Jones, just south-east of Flint, shows a road which might have been part of the Roman coastal route. On the *Deva-Segontium* itinerary, the intermediate towns of *Kanovium* and *Varae* have been identified as Caerhun and St Asaph, because their distances from Caernarvon, Chester and each other approximate to the calculated average between *mansiones* of 25 Roman miles.

Using the same argument the Pentre Farm complex is unlikely to have been a *mansio* because it is only 11 miles from Chester. Its identification as the site of a mansio, and home of *Varae*, would leave unequal distances for the three stages known to have existed between Chester and Caernarvon. That Pentre Farm was a *mutatio* or smaller counterpart of the *mansio*, acting as a relay station is also a possibility, but there was no evidence within the excavation area at Pentre Farm of any form of stabling. Thus Pentre Farm probably had nothing to do with the *cursus publicus*. If any category within Bender's general class of buildings called *mansiones* or *praetoria* fits Pentre Farm then it is his 'Rasthouser bei Lagern und Kastellen', which emphasises official connections without necessarily implying the continuous presence of soldiers.

It is yet another matter to assume that Pentre Farm lay close to a fort. If it did, it might have housed officials of the procurator's department as well as of the military at various times, perhaps even at the same time, since it is possible to interpret parts of the Pentre Farm ground plans as suites of rooms of comparable status.

There is no archaeological evidence to say whether

Pentre Farm was near a fort, but an historical appraisal gives an indication. Since Tacitus recorded that there was resistance from the Deceangli in AD 48 (Nash-Williams 1969), it may be assumed that some form of, as yet undetected, fort was erected in or near their territory at that time, before the withdrawal to cope with the rebellion of the Brigantes. By AD 60, however, the Deceangli had probably acquiesced. Certainly, no resistance was recorded during the drive towards Anglesey. The absence of known forts in Deceanglian territory has been taken as an indication of a peaceful, pro-Roman native population, and in these circumstances North Wales could have been adequately policed from the legionary base at Chester. The local lead-mining industry, on the other hand, might in itself have been considered enough of a prize to warrant guarding. It is known that exploitation must have been started by AD 70, from the date of the earliest lead pig, and at around this time a fort or forts may have been built to safeguard the newly-found economic potential. Soon afterwards, however, the situation changed with the appearance of imperial lessees like C. Nipius Ascanius whose name occurs on a pig of AD 74 (Webster 1953).

With civilian officials becoming increasingly involved, the need for direct military supervision would have decreased, so that by the time the Pentre Ffwrndan industrial settlement was established in about AD 90 the need for a fort might have gone altogether. Even if the pottery used to the date the Pentre Ffwrndan settlement is pushed back to the beginning of its range (about AD 90, p 120), it is impossible to see that site flourishing in time to produce, or be associated in any way with the known lead pigs. The earliest centre of production or processing must therefore have been elsewhere in the territory of the Deceangli and the Pentre Ffrwndan settlement must be separate from any conjectured but still occupied early fort.

The Pentre Farm complex may not, therefore, have been built alongside a fort at all; this is not to say that military interests would be entirely absent. Indeed, because all mineral rights belonged to the emperor some kind of military or official presence is likely; though it may have been managerial and concerned with the revenue for leaseholdings, rather than coercive and concerned with policing. As such it would not necessarily have needed a fort in the immediate vicinity: 'an officer or two - *centurion* or *optio* called on for technical direction, for example, would not have lived in a defended building' (MacMullan 1964).

Two apparently undefended sites in Cheshire, with official connections, may also be referred to in support of the idea that Pentre Farm, as an official building, need not imply the nearby presence of a fort. The twentieth Legion works depot at Holt stood on ground sloping gently down the fifty-foot contour to the edge of the west bank of the River Dee. The barracks had a large wall around them which would not have been sufficient for defence and the officer's house lay to one side of the compound, apparently without any enclosure at all. An industrial site at Heronbridge, similar to Holt in its elevation and relation to the River Dee, has been interpreted as the trans-shipment point of goods coming down-river from Holt *en route* for Chester and yet the site was not enclosed in any way (Thompson 1965).

Interpretation as an official building

To consider the site in its regional environment it matters little whether its correct interpretation is *mansio* or *praetorium*, only that it was official. One possible reason for such a building is that there were local port facilities at the mouth of the brook that runs beside Ship Field into the Dee. Until earlier this century there was a wharf at the mouth of this brook, according to local residents: although a wide expanse of marsh now separates the site of the Pentre Ffrwndan settlement from the tidal waters, this may not have been so in Roman times since shoal conditions seem to have deteriorated since then (Cleere 1978). But if port facilities are to be considered as the *raison d'etre* for a substantial building like that at Pentre Farm, the goods passing through must have been of greater importance than those required for an ordinary civil settlement.

One is forced to the most obvious conjecture, that the *raison d'etre* of both the Pentre Ffrwndan settlement and the Pentre Farm building is the lead mining industry. No direct connection between the industry and Pentre Farm can be shown, but one is strongly suggested by an assessment of the arguments which rely on general considerations rather than specific evidence from the Pentre Farm or Pentre Ffwrndan excavations.

The mineral wealth of Britain was one of the principle reasons for the Claudian invasion at a time when, according to Pliny, lead was 'mined with some difficulty in Spain and Gaul but in Britain it is present to such an extent near the surface that there is a law limiting its exploitation' (*Nat. Hist.* XXXIV 17, 164). North Wales contained rich superficial deposits of galena, especially those of the Prestatyn-Holywell series where the veins have been formed under a cover of shale (Webster 1953). Whether or not this ore was found to contain insufficient silver content for direct military exploitation is open to doubt, but there can be none about the demand generated by the legionary base and civilian settlement at Chester, where all kinds of fittings from pipes to coffins would have been made of lead. The earlier excavations at Pentre Ffwrndan demonstrated that some kind of lead processing was carried on from the late first to mid second centuries at least, regardless of whether or not this was actually smelting. A furnace packed with limestone chips and charcoal found in 1976, supports this idea and if furnaces discovered in 1840 (see p 3) are also taken into account, the total area covered by industrial features is considerable. Pentre Farm, built early in the Hadrianic period when lead processing still continued at Pentre Ffwrndan, shows no direct or neccessary connection with large-scale leadworking. The lead residue (oxide ?) used as floor materials in Stages IV-VIII suggests that processing may have continued nearby until late in the life of the complex. The only evidence of lead working at Pentre Farm is a small hearth in Stage III and an only tentatively identified cupellation hearth belonging to Stage VIII. It is difficult, however, on general rather than specific grounds, to avoid making the connection between Pentre Farm and the local lead industry. In this context, it is legitimate to speculate on the possible function of the Pentre Farm complex within the organisation of the local lead mining industry.

Function of the complex within the local lead mining industry

Who occupied the complex? The status of the complex itself is difficult to ascertain, but the floors and decorations of the better rooms in it compare with the materials found in centurions' quarters at Chester. The plan resembles those of tribunes' houses in legionary fortresses or commandants' houses in auxiliary forts. Someone of equestrian rank may be considered. Later in the sequence, from Stage IV onwards, with an apparent decline in quality of materials and precision of layout it is less easy to speculate in this fashion. The need for accommodation is obvious, since even if military officers might be lodged in the commandant's house, some provision must have been required for civil officials, not least the staff of the *procurator provincae* (Salway 1958). The administration of mines was under the control of the procurator who was responsible to the emperor for the management of imperial estates and all financial administration (Webster 1953).

Here a single freedman procurator was controller of those mines which were not important enough to warrant both an equestrian and a freedman procurator (Weaver 1965). Ulpius Aelianus was one such freedman and it was under his jurisdiction that the second of the Alljustrel plaques was issued. Although it is customary to assume that mining work in Roman times was carried out by slave labour, the regulations issued under Ulpius Aelianus, seem to suggest that while slave labour might be employed by individuals there was a wide liberty of action for *coloni* or tenants, and the regulations governing tenure seemed to be specially designed to encourage individual initiative (Allen 1970). It is in the implementation and supervision of these regulations that the need for procuratorial officials is found and the lead pig bearing C. Nipi Ascani (Webster 1953) testifies to the presence of lessees in North-east Wales from at least AD 74: perhaps a similar administration should be envisaged here. In the Alljustrel inscription the procurator at Vipasca is authorised to fine the lessee of the bath monopoly if he fails to provide the service required under his contract. He may also beat a slave and have him sold, or confiscate the property of a free man and exclude him from the mining districts if he is found guilty of theft or of contravention of various mining regulations. In addition, he may have to decide issues in court, such as breaches of monopolies sold by the procurator or disputes between partners in a mining concession (Brunt 1966). None of these powers is extensive but they do show that on the mining estates, procurators were vested with the same kind of powers as municipal magistrates. The Pentre Farm complex would have been adequate for such an official and its location near the place of processing and possible trans-shipment ideal for the execution of procutorial duties. The construction of the complex for such a procutorial official, *ab initio*, would also explain the military being very close, as the Mihilici inscription shows.

Alternatively, one might speculate on the presence of a military official at Pentre Farm. MacMullen (1964) has considered the administrative role of the military and explains that Egypt offers important information that can be used with confidence, since it is often confirmed by what is known of other provinces. Many of the soldiers in Egypt were based in camps or civil offices from which they were sent to do their administrative jobs. Others were scattered in minor posts, *stationarii* in stationes. Outside Egypt *stationarii* are met with frequently, sometimes under the name of *regionarius*, to indicate responsibility for a rural district. *Stationes* were held by detachments living in their own tents, in private houses or in unfortified *mansiones* (MacMullen 1964). The last comment may make such military officials seem appropriate for Pentre Farm, but the duties of the *stationarii* described by MacMullen are hardly relevant to a lead-mining district. These ranged from chasing criminals, to guarding the governor's residence or overseeing the confiscation of property. Typically they were attached to roads and communication centres, to keep the peace and help with movement of draught animals, drovers and the *annona*. In contrast to these duties those of the procuratorial department described earlier in the disscussion seem better to characterise the authority and responsibility of the occupant of Pentre Farm.

The complex as part of an Imperial Estate

Closely bound up with the presence of procuratorial officials is the implication that the area in which they worked was an imperial estate. These certainly existed in Britain, and an inscription from Combe Down, Bath (RIB 179), set up in the reign of Caracalla to record the restoration of a ruined headquarters by Naevius, imperial freedman and assistant to the procurators, records an imperial interest, possibly in the local Bath Stone (Frere 1974). In southern Britain there are two very large areas where the absence of villas and even towns is noticable. The first is Salisbury Plain and Cranbourne Chase where an area of over 900 square miles is virtually devoid of villas, although farmsteads and villages of 'peasant type' abound (Frere 1974), or in the Fen basin where an area three times as large has drainage works on a massive scale which imply official action. It is probable that the territories were public land, peasant settlers receiving farms in return for rent, and no doubt being placed under the control of a *procurator saltus* (Frere 1974).

Since the territory of the Deceangli in which Pentre Farm lay seems similarly devoid of town and villa settlement, and bearing in mind that mines were imperial property anyway, it is possible that some kind of imperial estate existed here. The exclusive distribution of native homesteads in north-west Wales seems to coincide directly with an area of intensive military presence. Settlement in north-west Wales, on the other hand, perhaps continued as before, in an essentially pastoral economy (Kelly 1977). In the territory of the Deceangli, this might have been as a result of the confiscation of the areas around the mines. 'In Galatia the procurator is at least associated with the legate in determining the boundaries between the city of Sagalassus and the village of Tymbrianissus. Even though there is no direct testimony that the village was in an imperial estate, I feel that the procurator's presence must be explained on the hypothesis that it was; it was the fiscal interest that brought him in...' (Brunt 1966).

Status of the Roman Lead Mining District of North-east Wales

If the Pentre Farm complex is to be seen as an official building integrated with the lead mining industry, some consideration of the status of that industry seems appropriate. This has been done by assessing the Flintshire mines together with the most successful Roman mines in Britain on the Mendips. The opencast Iron Age mines of the Mendips were harnessed by the Romans soon after the conquest and a fort was established in about AD 45 to protect the mines of high-quality ore. A second fort may have been built in the second century. The scale of the Mendip industry and its high silver content suggests that a strong military presence remained even though lessees are known from the time of Nero. The Mendip industry was well placed for serving markets in the south of *Britannia* and in other provinces via *Sorviodorum* and *Clausentium* on the south coast. Mendip lead pigs have been found in St Valery and Lillebonne and in Pompeii, where isotopic analysis has revealed that the lead from a cistern may have been imported from the Mendips between AD 49 and 79. The Lillebonne pig is of the Severan period and testifies to the maintained importance of the Mendip industry. In the third century the industry received new impetus when it was used extensively in the manufacture of pewter and the large number of West-Country villas ensured that there was a good market.

The Flintshire mines, too, were possibly in existence in the Iron Age. Certainly they were being worked by the Romans very soon after the conquest of the area in AD 58-9. The appearance of lessees like Ascanius, however, in the reign of Domitian at the latest, suggests that direct military exploitation had ended, perhaps because the silver content of the ore was lower than anticipated. If this was so, imperial revenue could perhaps be equally well secured from levies on civilian leaseholders, without the risk of large-scale imperial investment. On available evidence, markets for Flintshire products never seem to have been as widespread as those of the Mendips. Chester must have used a large percentage of the lead produced and the farthest Flintshire pigs are known to have reached is Runcorn in Cheshire.

In the second century a situation can be envisaged in which civilian involvement in the mines had grown substantially or even taken over from the military altogether, and since the Pentre Farm complex probably dates initally from the period of Hadrianic stimulus to frontier development in the 120s, one may speculate that it derives from some internal reorganisation of the Flintshire mines. Both the Pentre Farm and Pentre Ffwrndan settlements generally, were apparently unaffected by the changes in the north of the province under Antoninus Pius and this too may indicate the continued importance of the Flintshire industry in the later second century. The changes evident in the Pentre Farm complex in Stage VIII of *c* AD 200 may mean that the formerly official complex had become a private structure, possibly including a workshop, though not necessarily one dealing with lead. How this relates to late second century Welsh history, when there is most uncertainty about pottery dating, is not possible to say. The Flintshire mines appear to have played no part in the pewter industry of the third century and the only other scattered finds from the Pentre Ffwrndan settlement are of mid-fourth century date.

The apparent demise of the Pentre Ffrwndan settlement may well be mirrored by that of a site at Municipium D. D. in Upper Moesia, where the management of the mines was in the hands of imperial authorities (procurators and freedmen), whose official position gave them little interest in the establishment of community life along urban lines (Mócsy 1970). The eventual effect on the isolated community was that it fell into decline as soon as the mines, on which its existence was based, ceased to function (Cerskov 1970).

Plate I: General view of excavations, looking south, showing foundation trenches for buildings of Stage II

Plate II: Yard 7A between the West Range (left) and South Range (right) with the North-west Range in the background. Looking north. Stage IV.

Plate III: Detail of the east wall of the West Range showing post-pipes and plaster face. Looking west. Stage IV.

Plate IV: General view of excavations, looking north-west. Stage VII.

Plate V: General view of Bath-house and part of Northwest Range, Stage VI and underlying West Range of Stage IV. Looking north.

Plate VI: Detail of Rooms 47 and 51 in the Bath-house, Stage VI. Looking north.

Plate VII: Detail of pier in Bath-house, Room 47, showing revised *pila*, Stage VI. Looking west.

Plate VIII: Detail of Bath-house drain F134, with lead pipe F131 feeding into it. Looking west. Stage VI.

Plate IX: Plunge bath F130, looking south-west, with lead pipe F131 in foreground. Stage VI.

Plate X: Plunge bath F130, looking east. Stage VI.

Plate XI: General view of excavations, looking west. Stage V.

Plate XII: Detail of F101 in Rooms 39 and 40 of the North Range, looking north. Stage V.

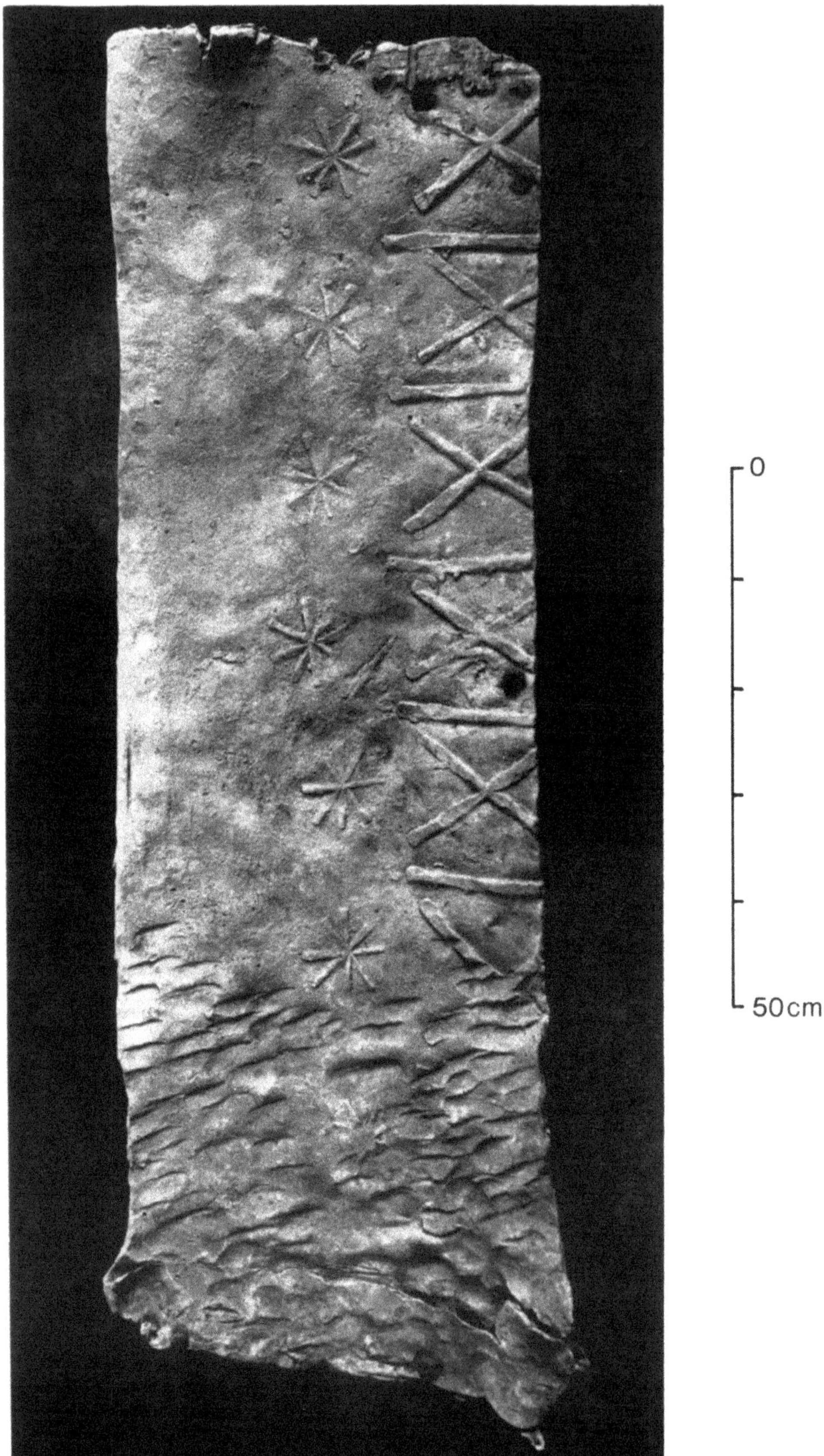

Plate XIII: Lead sheet, no 47. (Photograph, B V Willams)

Plate XIV: Lead junction, no 40

Plate XV: Stone with barnacles from Bath-house, Stage VII.

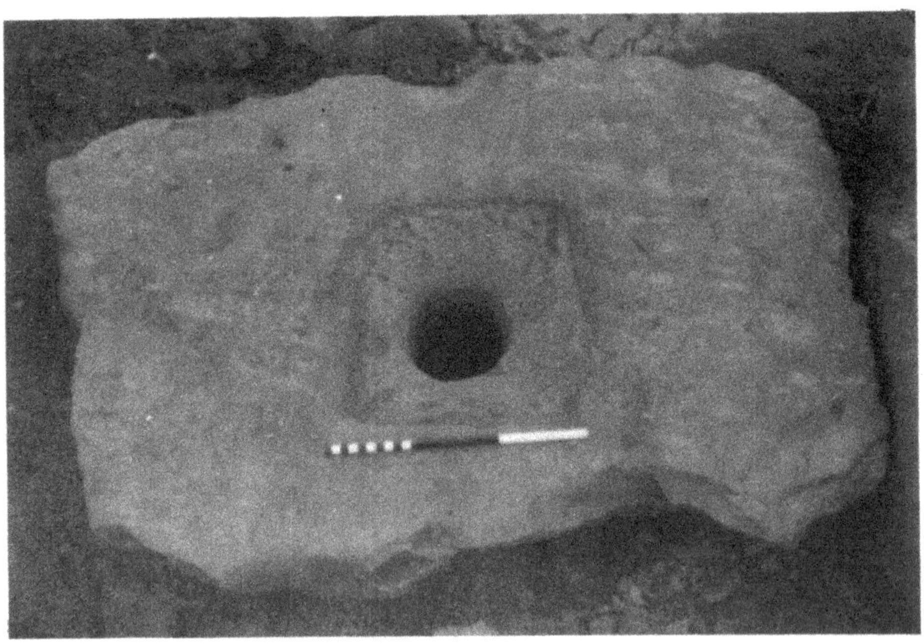

Plate XVI: Stone with drain hole, from Room 39, Stage V.

Part 2: The finds

INTRODUCTION (KB)

The majority of the small finds reports were written in 1976-7 and only slightly revised in 1987. They remain essentially the original reports.

Many of the finds have been mislaid and were not available for study in 1987, these include gold, copper-alloy and small lead objects, worked bone, animal and human bones and stone (the bench leg survives).

THE COINS

By P J Casey (1976)

The condition of the coins at the moment of deposition has been indicated by a notation of the wear, where this can be ascertained. The notation is subjective and has no chronological value:

Table 1: The coin list

Layer	SF No	Issuer	Denom	Type	Ref RIC	Date	Cond
1/1 (post-complex)	3	M.Antonius	Den	ANT AVG III VIR R P C LEG ...	Braw 544.8	32-31BC	EW/EW
3/97 (Stage IV)	396	Titus/Domitian Caes.	As	- S C - Spes	as 656	AD73-78	W/UW
10/93 (Stage III)	414	Vespasian	Dup	IMP CAES VESPASIAN AVG COS VIII PP FIDES PVBLICA	753(a)	Lugdunum AD77-78	W/W
3/94 (Stage VI)	372	Domitian	Sest	IMP CAES DIVI VE SPF DOMIT IAN AVG P M	233	AD81	UW/EW
12/60	269	Domitian	As	MONETA AVGVST	301(b)	AD85	W/W
1/60 (post-complex)	22	Trajan	As	IMP CAES NERV A TRAIAN AVG GERM P M TR POT COS II - S C	393	AD98-117	W/W
2/11 (Stage V)	126	Trajan	Sest	Illegible	-	AD101-117	UW/UW
12/52	399	Trajan	As	IMP CAES NERVAE TRAIANO AVG GER DAC PMTRP COS V PP SPQR OPTIMO PRINCIPA - SC	549	AD103-111	SW/cor.
1/60 (post-complex)	8	Hadrian	Den	IMP CAESAR TRAIAN HADRIANVS AVG P M TR COS II. PIETAS	45a	AD119-128	SW/UW
Topsoil	56	Hadrian	As	CLEMENTIA AVG COS III PP - SC	714d	AD120-138	UW/W
1/60 (post-complex)	4	Hadrian	Sest	HADRIANVS AVGVSTVS IVSTITAI AVG COS III PP - SC	710c	AD128-138	SW/SW corroded
Topsoil	90	Favstina I	Dup/As	AETERNITAS SC	1157	AD141-161	UW/UW
2/11 (Stage V)	130	Ant Pius	As	IMPERATOR II SC	728	AD143-144	UW/UW
1/64 (post-complex)	24	Favstina II	Den	FAVSTINA AVGVSTA HILARITAS	606	AD161-176	SW/SW corroded

UW: Unworn: suggests a virtually uncirculated coin; SW: Slightly worn: highest parts of the design a little flattened by wear; W : Worn: design well abraded but legends visible; EW: Extremely worn; illegible.

Of the fourteen coins recovered from the excavations only five were from stratified levels within the complex, the remainder being residual or in post-complex deposits. Those in stratified contexts were also apparently residual since the latest coin (AD 143-44 from Stage V) was found in a late second to early third century context.

In general the coins support the pottery dating, indicating a starting date for occupation in the later first century AD. Further interpretation from the small number of coins found would be unwise. KB

OBJECTS OF GOLD

By Janet Webster (1976-7)

1. (Not illustrated). Small fragment of gold leaf. (SF 5, topsoil above the complex).

OBJECTS OF COPPER ALLOY (FIGS 21-3)

By Janet Webster (1976-7)

1. The lower half of a trumpet brooch. (SF 30, post-Roman/64). The break occurs at the top of the waist knob. The button is flanked by plain mouldings and all the extant mouldings are carried round the back of the brooch. The leg is plain. There is an elaborate foot consisting of four mouldings and terminating in a projecting knob. Trumpet brooches are known to have been in circulation at least as early as the late Neronian or early Flavian period (J Webster forthcoming b; Hobley 1969, 110, fig 19, no 9) and it is suggested that their evolution may be even earlier (Boon and Savory 1975, 41-61). The trumpet brooch maintained its popularity over a considerable period and the fully evolved type with acanthus waist moulding was still flourishing in the Antonine period (Cunliffe 1968, 81, pl XXVIII, no 30).
2. A disc brooch of the type often known as a 'buckler'. (SF 120, 2/7, Room 51, Stage VII). The brooch is hinged. The disc has a marginal rim and six projecting lugs. The centre of the disc rises to a cupped stud with a small central boss. There are two concentric incised circles just within the marginal rib.
   The type is briefly discussed by Hull (1967, 56, fig 22, no 207) and Butcher (1974, 128, fig 55, no 29) where the occurrence of the type in Pannonia, Germany and Belgium is noted, and a number of British parallels are cited. There seems to be little firm dating but the second and third centuries are suggested.
3. Bronze brooch with a hinged pin. (SF 428, 11/33, Room 12, Stage III). The crossbar for the pivot is constructed so that from the front it has every appearance of being the side wings of a sprung brooch while at the back the casing is recessed within the lines of the 'side wings'. The pivot bar projects through the end of its cover at one side much as the axial bar of a sprung-pin brooch might project through pierced discoid terminals to side wings. The crossbar is elaborately moulded on the front only and there are small semi-circular mouldings to either side of the head. In profile the bow is sharply angular at the head and then describes a gentle curve. At the top of the head is moulded a V-shaped tongue set within a plain rectangular panel bordered by two outer ribs which extend as far as the enamelled panel, where they merge into the outer borders. The enamel decoration extends all the way down the front of the bow. It consists of two bands of alternating blocks of blue and yellow enamel with a bronze wall down the centre. At the top of the panel two elongated V-shaped mouldings of bronze interrupt the first blocks of colour. The tapering bow terminates in a small knobbed foot. Brooches of this type are developments from the Polden Hill with Dolphin profile series (J Webster forthcoming b, nos 24f). The profile changed from the hump at the head and flattish bow to a sharp angle and graceful curve. The pin may be sprung or hinged, if hinged the appearance of side wings is retained; the semi-circular mouldings to either side of the head of the bow are retained. The Pentre Farm example may be compared with several brooches from Wilderspool; one (Webster J forthcoming a, no 1) has a sprung pin and a decorated but not enamelled upper bow; another, a brooch in Warrington Museum (Thompson 1965, 82-3, fig 20, no 4), has the same treatment of the housing for the pivot of the hinged pin with, apparently, elaborate side wings to the front and in fact a recessed casing forming the crossbar to the rear. The brooch differs from the Pentre Farm one in having flanged mouldings down the sides of the bow and in the fact that the enamel decoration extends only half-way down the bow. In view of the Wilderspool parallels it seems reasonable to suggest a second century date for the Pentre Farm brooch. Similar brooches have recently been found at Prestatyn (Mackreth, forthcoming, fig 36, nos 5-7).
4. The hinged lid of a bronze jug. (SF 92, 0/20, Room 11, Stage III). The lid is approximately heart-shaped but the lower end of the heart, instead of ending in a point, is elongated and cut off straight, no doubt to accommodate the spout of the jug (Tassinari 1975, pl XXXV, no 180). The heavy medial rib and projection from it remain on the top of the lid. The projection

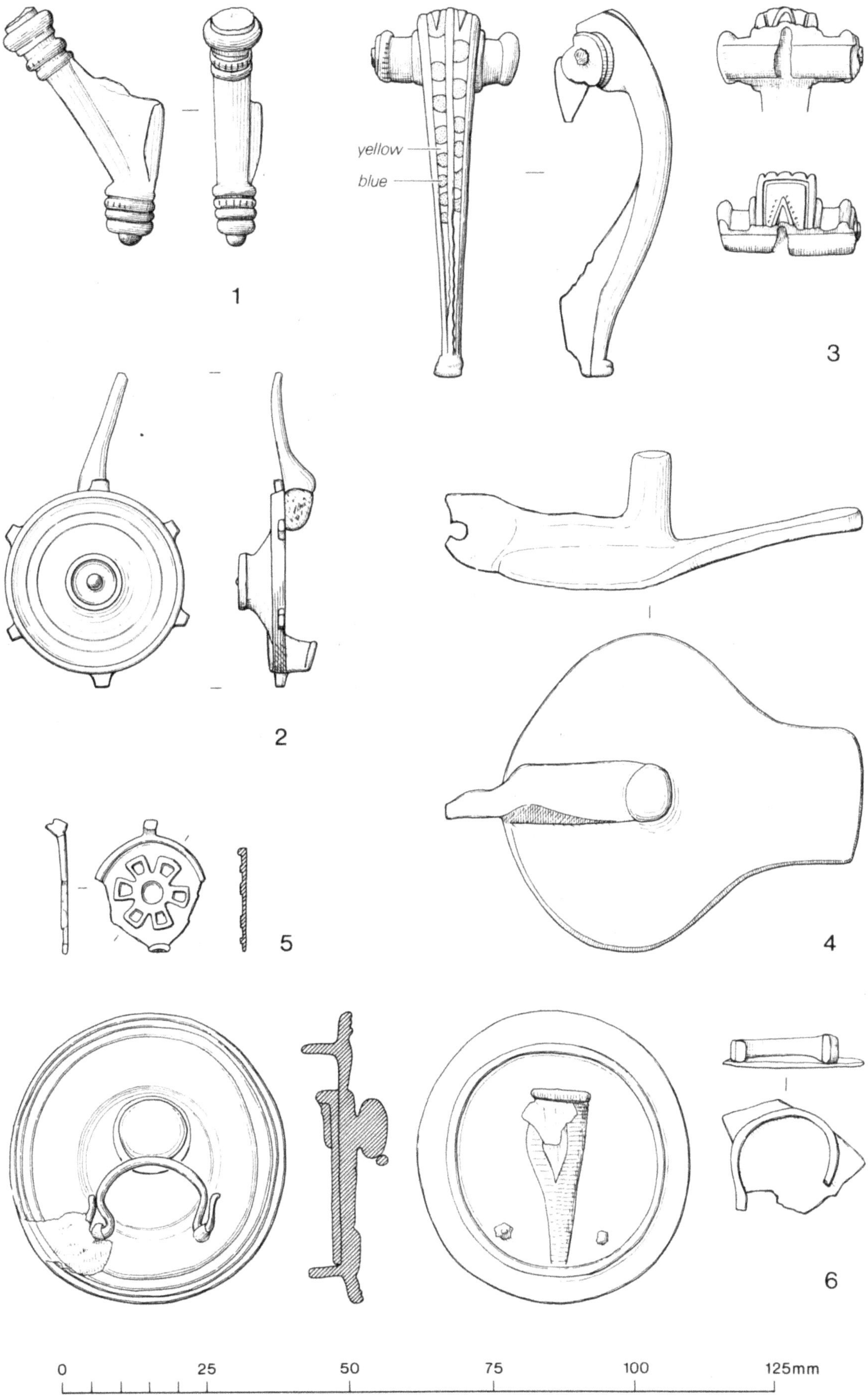

Figure 21: Objects of copper-alloy, full size

probably ended in a knob by means of which the lid could be lifted. Traces of the hinge remain at the outer end of the rib (den Boesterd 1956, pl XI, no 249; 70, nos 247-53, dated 1st to 3rd century AD).

5. Seal-box lid. (SF 56, post-Roman/60). The object is damaged but the lid must have been approximately oval in shape. There is a ridged margin with, at the top, the pierced flattened lug which formed the central leaf of the hinge. The lid is decorated with a six-petalled flower defined by bronze walls. Each petal and the circular centre were originally inlaid with enamel. A number of Pannonian leaf-shaped seal-box lids with flower motifs have been published (Sellye 1939, 84, pl XV, nos 12-13; nos 14-15). See also an example from Chester (Lloyd-Morgan 1978, 31, fig 26, no 7).
6. The lid of a bronze inkwell. (SF 367, 6/54, Room 29, Stage V). See Comstock & Vermeule (1971, 324, nos 454-5) for complete examples of inkwells of this type. The Pentre Farm lid is equipped with a knob and with a small loop handle on top. On the underside of the lid, a short way from the edge, is a downward-projecting flange which describes a full circle and presumably sat within the aperture of the body of the inkwell and served to give a secure fit. The rivets which secure the loop handle protrude through to the back within the circle defined by the flange. The shaft of the knob passes through the lid and serves to secure a bronze or iron tongue of unknown purpose; this cannot have served as a catch as its point will have been inhibited by the flange. There are a number of ?iron fragments associated with the inkwell top, one of which has a projecting bronze flange, now describing a small incomplete circle. These fragments may have constituted a lining to the underside of the inkpot lid, to conceal the methods of fixing the knob and handle. If so, the small circular flange of the lining seems to indicate an appropriately small aperture for the removal of ink.
7. Candlestick. (SF 260, 0/48, post-complex). The object comprises two bell-shaped cups, back to back, with a decorative moulding between. One cup has an internal spike centrally placed, the other has a small circular recess in the same position.
   The object should be compared with a larger and slightly more elaborate example from Heddernheim (Fischer 1973, 107-8, fig 27, no 1) which retains the cylindrical housing for a candle in one of its cups and evidence of the same housing in the other. Fischer lists a number of Continental parallels. The Heddernheim candlestick was clearly reversible as is the Pentre Farm example, though the advantage of this is not obvious. Because of the small size of the cups the Pentre Farm candlestick would not have needed internal cylinders to house the candles, as the latter would have been adequately held by the edges of the cups, especially with the additional stability afforded by a spike. A similar reversible candlestick was found at Brough, East Yorkshire (Corder 1935, 32, fig 8, no 6) where other examples from Britain are cited, namely one from Castle Hill, Whitton, Ipswich, and a single cup, perhaps from a similar double candlestick, from Lincoln and now in the British Museum (I am most grateful to I Scott and S Thompson for drawing the Heddernheim reference to my attention).
8. Small plumb-bob. (SF 415, 10/100, midden outside Rooms 24-25, Stages II-III). The projection at the bottom of the bob is recessed perhaps for a pointer. The knob at the top of the bob is pierced for attaching to a line.
9. Bronze lock-plate. (SF 342, 6/51, Room 29, Stage IV).
10. Small tumbler lock bolt. (SF 34, post-Roman/64).
    The workings of this type of bolt with the tumbler lock slide-key are described by Wheeler (1930, 70-71, fig 16). The Pentre Farm item must have been used to fasten a box like the bolt from Finsbury Circus (illustrated by Wheeler 1930, 73, fig 17, no 2), to which it is similar in size though the piercing is not so complete.
11. Bell-shaped stud. (SF 34, 6/24, demolition between Stages III-IV. Probably derived from Room 24 or 25).
    The fitting has a broad shaft of bronze cast in one with the knob. One edge of the shaft is carefully shaped but the other is roughly finished save for a V-shaped nick towards the outer edge. The end of the shaft is damaged.
    Studs such as this have recently formed the subject of a study by Allason-Jones (in Bishop 1985, 95-108). The Pentre Farm fitting is of Allason-Jones's Type 2, characterised by its long bronze shank, twice as wide as it is thick. She suggests that this type of bell-shaped stud was used for attaching decorative panels or lock-plates to boxes or chests and is to be found on both military and civilian sites.
12. A Bell-shaped stud similar to the last but with an iron shank and, therefore, of Allason-Jones's Type 1. She suggests that these studs were used for a variety of purposes and notes a predominantly military distribution. (SF 2, topsoil above complex).
13. Heavy functional ring. (SF 17, post-Roman/60). Incomplete.
14. Similar. (SF 55, topsoil above complex).
15. Functional ring. (SF 369, 6/58, Room 29, Stage IV).
16. Functional ring. (SF 339, 13/85, yard to the

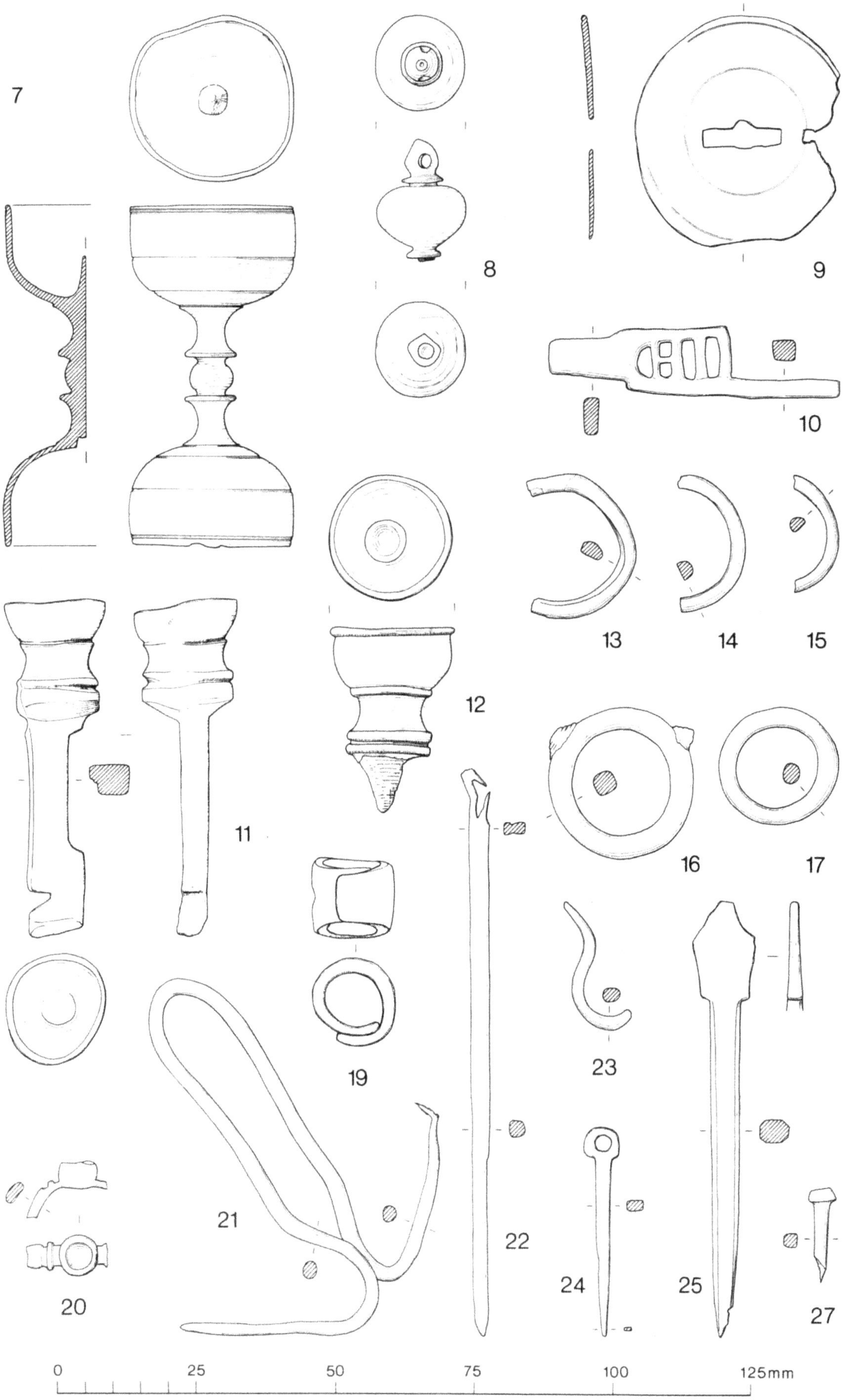

Figure 22: Objects of copper-alloy, full size

north of the complex, Stage V).

17. Functional ring. (SF 295, 13/60, yard to the north of the complex, Stage V).
18. (Not illustrated). Fragments of a thin functional (?) ring with a sliver of (?) leather associated. (SF 79, topsoil above complex).
19. Ferrule. (SF 318, 6/18, Room 30, Stage IV). For similar objects see (Cunliffe 1971, 124, fig 53, nos 180-81; Brailsford 1962, 1-2, fig 1, nos 18, A19; Ritterling 1912, pl XVI, nos 28-33; Hobley 1974, fig 24, no 43), and examples from Usk (J Webster forthcoming b).
20. Fragments of a finger ring. (SF 273, 2/30, Room 51, Stage VI). The bezel comprises a circular bronze wall which once housed a semi-precious stone, enamel or glass setting. The shoulders are decoratively ridged and notched. A more elaborate example from Chesterford, Essex is given by Marshall (1907, 216, fig 155, no 1406; also Butcher 1967, 22, fig 8, nos 6-13; where a second century date is suggested).
21. Bronze double hook, perhaps for use on a steelyard or balance. (SF 44, topsoil above complex). There are examples from the Walbrook, London (British Museum 1922, 75, fig 40, no 11) and from near the Mansion House, London (Wheeler 1930, 85, fig 22, no 2).
22. Bronze needle, now damaged at the eye. (SF 277, 2/37, Room 51A, Stage V).
23. Small bronze hook. (SF 42, topsoil above complex). Perhaps a fish hook.
24. Probably a single-pronged nail cleaner. (SF 135, 0/8, post-complex). As it is pierced at the head for suspension it could have been part of a toilet set. (Cunliffe 1975, 202, fig 110, no 20; there wrongly identified).
25. Stylus. (SF 19, topsoil above the complex). The point is lost.
26. (Not illustrated). Part of a brooch spring, comprising part of the coiled wire and axial bar. (SF 29, post-complex/64). Probably from a late first or early second century brooch.
27. Small bronze nail or tack. (SF 77, 0/1 post-complex).
28. Tiny expanding bracelet for a baby, or perhaps a hair-ring. (SF 65, topsoil above the complex). The overlapping terminals are tapered to accommodate each other. The ring is plain.
29. Ornamental foot. (SF 331, 1/55, Room 7, Stage VI). There is a projecting shaft for attachment behind and a rolled foot at one end; the opposite end is broken. The foot was probably attached to some small wooden object.
30. Bronze fitting. (SF 373, 6/24, demolition between Stages III and IV. Probably derived from Room 24 or 25). Like the previous piece there is a projecting shaft to the rear of the object for purposes of attachment. The fitting is broken at both ends.
31. Either a pin with a knobbed head and a small length of shaft or, more likely an ornamental nail. (SF 275, 2/32, Room 51, Stage VI).
32. Fragments of bronze chain, comprising one oval loop, open-ended, and a fragment of another, and one circular loop with ends over-lapped and pressed together. (SF 270, 2/33, Room 51a, Stage V). Although there are marked wear facets on each of the links, it may be that the circular loop represents a repair of the chain and is a replacement.
33. (Not illustrated). Perhaps a fragment of a link of chain. Similar to the oval loops described above (no 32) though more angular. (SF 23, post-Roman/60).
34. Silvered bronze cone, perhaps from the centre of a disc brooch. (SF 321, post-Roman/17).
35. Possibly a crude finger ring. (SF 188, post-Roman/60). Otherwise a reinforcement or fastening like a staple.
36. Small decorative stud? (SF 272, 2/32, Room 51, Stage VI). The object is rectangular and slightly curved.
37. Stud with flange on the underside, pierced by a flat headed nail which passes through an onion-shaped washer on the underside of the stud, cf two studs from Fishbourne (Cunliffe 1971, 124, fig 53, nos 182-3). Like the Fishbourne examples the Pentre Farm stud also retains traces of lead. (SF 14, post-Roman/60).
38. Small object, bell-like in appearance, with cut-away sections to either side. (SF 338, 6/47, Room 29, Stage V).
39. Lozenge-shaped fragment of sheet bronze with the ends folded towards the centre and bent back out again. (SF 279, 2/36, Room 51, Stage VI). Neal (1974, 134, fig 59, nos 104-106) suggested that such objects served as improvised split pins or rivets for leather work.
40. Thin binding. (SF 1, topsoil above complex).
41. (Not illustrated). Squashed domed stud. (SF 334, 6/10, Room 30, Stage VII).
42. (Not illustrated). Sheet bronze fragments. (SF 343, 6/54, Room 29, Stage V).
43. (Not illustrated). Small fragment of domed stud. (SF 49, topsoil above the complex).
44. (Not illustrated). Fragment of pin. (SF 355, 7/89, Yard 8, Stage I).
45. (Not illustrated). Fragments of thin bronze binding. (SF 235, 1/41, Yard 46, Stage VII).

NON-ROMAN FINDS OF COPPER-ALLOY (NOT ILLUSTRATED)

By Martin Henig (1976)

1. Buckle, strap-end type. D-shaped with tongue 16 mm. The attachment plates survive in a

somewhat corroded state (*c* 18 mm) but the pin is missing. (SF 37, topsoil, above medieval robbing of bath-house).
Mid fourteenth century *London Mus Med Cat.* 1940, 272, pl LXXV, no 8; Biddle, Barfield & Millard, 1959, 182, fig 19, no 4, stratified AD 1350-1426).

2. Buckle, stirrup shaped. Pin missing 31 mm. (SF 27, topsoil, north part site).
The buckle is paralleled by two similar examples in the Ashmolean Museum Acc, nos 1927, 6272; 1927, 6304; both from Dunwich

3. Pin with ovoid head pierced; 32 mm perhaps for suspension, probably medieval. (SF 9, topsoil above medieval robbing of bath-house).

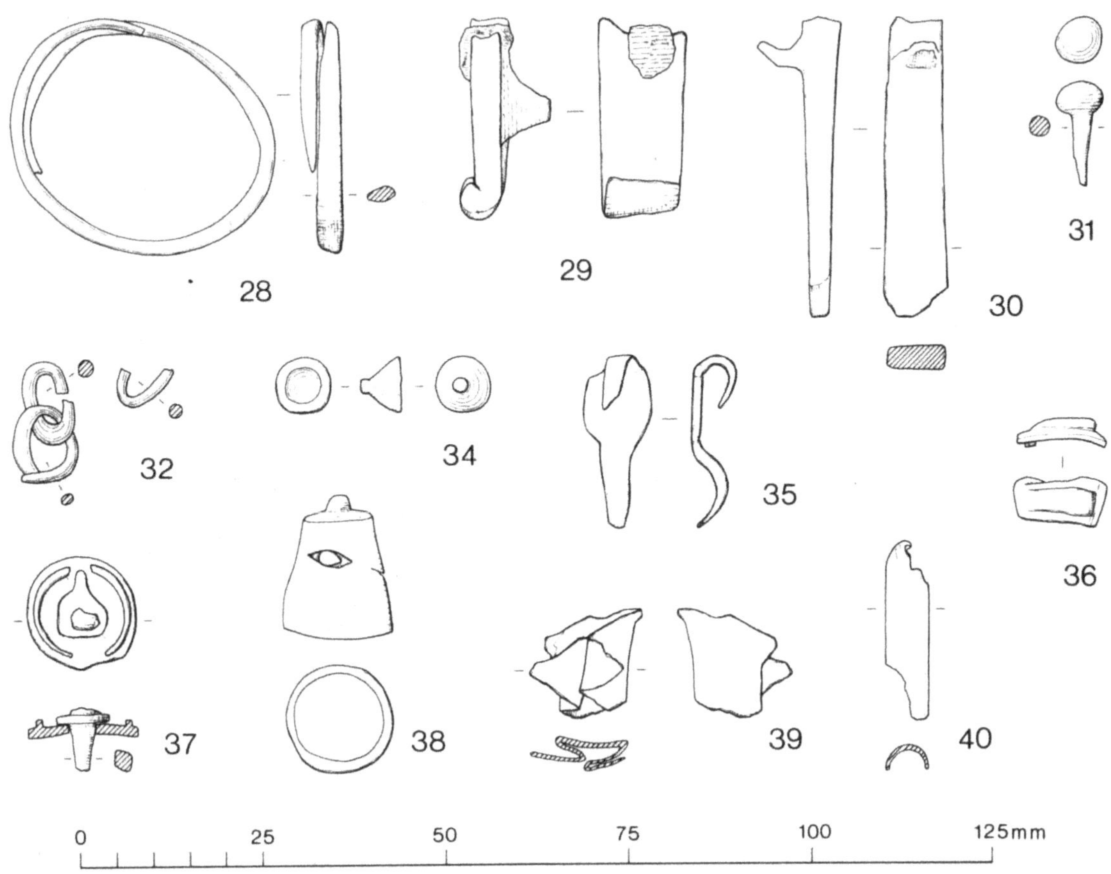

Figure 23: Objects of copper-alloy, full size

OBJECTS OF LEAD (FIG 24)

By Janet Webster (1976-7)

1-5. A series of leech-like runnels of lead. (SF 432, from Manchester University Excavation 1976, late Stage VIII). Several have been cut off with shears at the broad end. The objects have not been cast, as their rough finish shows no sign of casting flashes and they cannot, therefore, represent trial runs for brooches. The objects are runnels or overflow drips of lead which have been trimmed from larger items. (I am most grateful to Professor W H Manning for discussing these objects with me and for making many valuable comments).

6-7. Lead weights, 3.5 lbs, 1.60 kg and 6.25 lbs, 2.80 kg respectively. (SF 298, 9/32, Room 39, alterations to Stage V).

8. (Not illustrated)Large lead disc with deep circular recess close to one edge. (SF 51, topsoil above the complex).

9. Curved strip of lead binding. (SF 183, 0/35, post-complex). Two original edges remain and describe a rectangular corner. Three iron nails with large heads are in evidence along the opposite long edge.

10. Heavy lead sheathing. (SF 329, 13/33, yard to the north of the complex, Stage V).

11. (Not illustrated). Similar object, narrower and more cylindrical. (SF 319, 6/46, Room 29, Stage VII).

12. Lead off-cut. (SF 433, 10/96, Room 24, Stage II).

13. Small ring of lead, presumably for binding or sealing. (SF 310 13/67, yard to the north of the complex, Stage V).

14-15. Twisted rods of lead. (SF 435, 13/8, Stage VI or later, and SF 177, topsoil above the complex).

16. Large ?ring of lead. (SF 434, post-complex/64).
17. (Not illustrated). Lead spud. (SF 431, 0/1, demolition between Stages VII and VIII).
18. (Not illustrated). Bent rectangle of thick sheet. (SF 43, 0/1, demolition between Stages VII and VIII).
19-26. (Not illustrated). Melted amorphous pieces. Eight items from various contexts.
27-35. (Not illustrated). Strips of various shapes and cross-sections. Nine items from various contexts.

LEAD WATER-PIPES AND SHEETS

By Kevin Blockley and Keith Maude

Water-pipes (Fig 25, pl XIV)

Nine lengths of lead water-pipes were retained from the excavations; four from trial work conducted in 1976 by Professor G D B Jones (nos 41-44) and five from the main 1976-77 excavations (nos 36-40).

Pipes 36-40 were from F29 in the Stage II courtyard and the remainder from the Bath-house. Lead pipes F100, from the North Range in Stage V, and F131, from the Bath-house in Stage VI, were not retained for study.

36. Length of water-pipe, shield-shaped cross-section. Length 0.17 m.
37. Length of water-pipe, shield-shaped cross-section. Length 0.67 m.
38. Length of water-pipe, shield-shaped cross-section. Length 0.82 m.
39. Two lengths of water-pipe with shield-shaped cross-sections, joined by a lead junction box 110 mm long. One end of the pipe is not broken and may well represent an outlet. Length 0.5 m.
40. Two lengths of water-pipe with shield-shaped cross-sections, joined by a lead junction box 90 mm long (pl XIV). One length of pipe is complete, measuring 1.6 m long and shows signs of a further junction box (only a stain remains) at its open end. Overall length 2.2 m.
    The pipes above (nos 36-40) were manufactured from sheets of lead 5-7 mm thick, folded round so that the edges were brought closely together and the joint sealed by a run of molten lead (perhaps poured into a mould along the joint seam) which was then hammered down on cooling (Fig 25, nos 39-40). Cross-sections measured 40-50 mm by 60-77 mm.

The junction boxes were cast in an inverted position in the form of an inverted Y. To join two lengths of water-pipe the ends of the pipes were placed in-between the forks of the Y, the ends folded round the pipes and the two ends sealed with molten lead. Ribbing can be seen on the sides of both junctions, perhaps to give extra strength (pl XIV). The flat top of the casting, once inverted into its functional position, would have served as a support for the water-pipe.

41. Length of water-pipe with ovoid cross-section 85 mm by 65 mm, 7 mm thick. Length 0.85 m. The lead sheet, from which the pipe was fabricated, has been joined along the seam by pouring molten lead into a mould along the seam and this cast lead joint is very apparent (Fig 25, no 41).
    One end of the pipe has been belled out, perhaps in order to allow another length of pipe to be inserted, this belling also demonstrates the poor quality of the joint, in that one side of the joint has been opened rather than the metal stretching.
42. Length of water-pipe with shield-shaped cross-section 70 mm long axis, 40 mm short axis, 5 mm thick. Length 0.8 m. The joint has been fused by soldering or by cast metal and the joint has then been tidied up by means of hammering (as nos 36-40) presumably as a means to consolidate the joint.
43. Length of water-pipe with a flattened circular cross-section 55 mm by 50 mm. Length 0.45 m. As in no 42 the joint has been fused by soldering or by cast metal, but in this case the joint has not been hammered (Fig 25, no 43).
44. Length of water-pipe with ovoid cross-section 120 mm by 80 mm. Length 1.25 m. At one end it has been cut and opened out into a shovel shape and in the case of this pipe the joint has not been fused. The probable use of the pipe is that of a discharge outlet.

Sheet lead (pl XIII)

All of the sheets were recovered from Burial 2 (p 47). Four lead sheets were recovered from the burial but only three were available for study.

45. Lead sheet 0.22 m by 0.50 m, 5 mm thick. No cut edges, beaten surface. Three iron nails are visible piercing the sheet around its circumferences, one surviving to a length of 30 mm.
46. Lead sheet 0.27 m by 0.90 m, 5 mm thick. No cut edges, beaten surface. Six iron nails are visible piercing the sheet around its circumference, one surviving to a length of 30 mm.
47. Decorated lead sheet 1.24 m long 0.41-0.42 m wide, 10 mm thick. Three edges appear to be original, but partly trimmed, whilst the fourth edge has been flattened by a large number of hammer blows (with a hammer head measuring *c* 50 mm by 30 mm). The sheet is pierced around its circumference by six iron nails.
    It is evident that the decoration on the sheet was produced by placing strips of lead in the base of the mould before pouring the molten lead into the mould. The design may be seen in pl XIII.
    The object may originally have been part of a water tank but not enough survives to be certain.

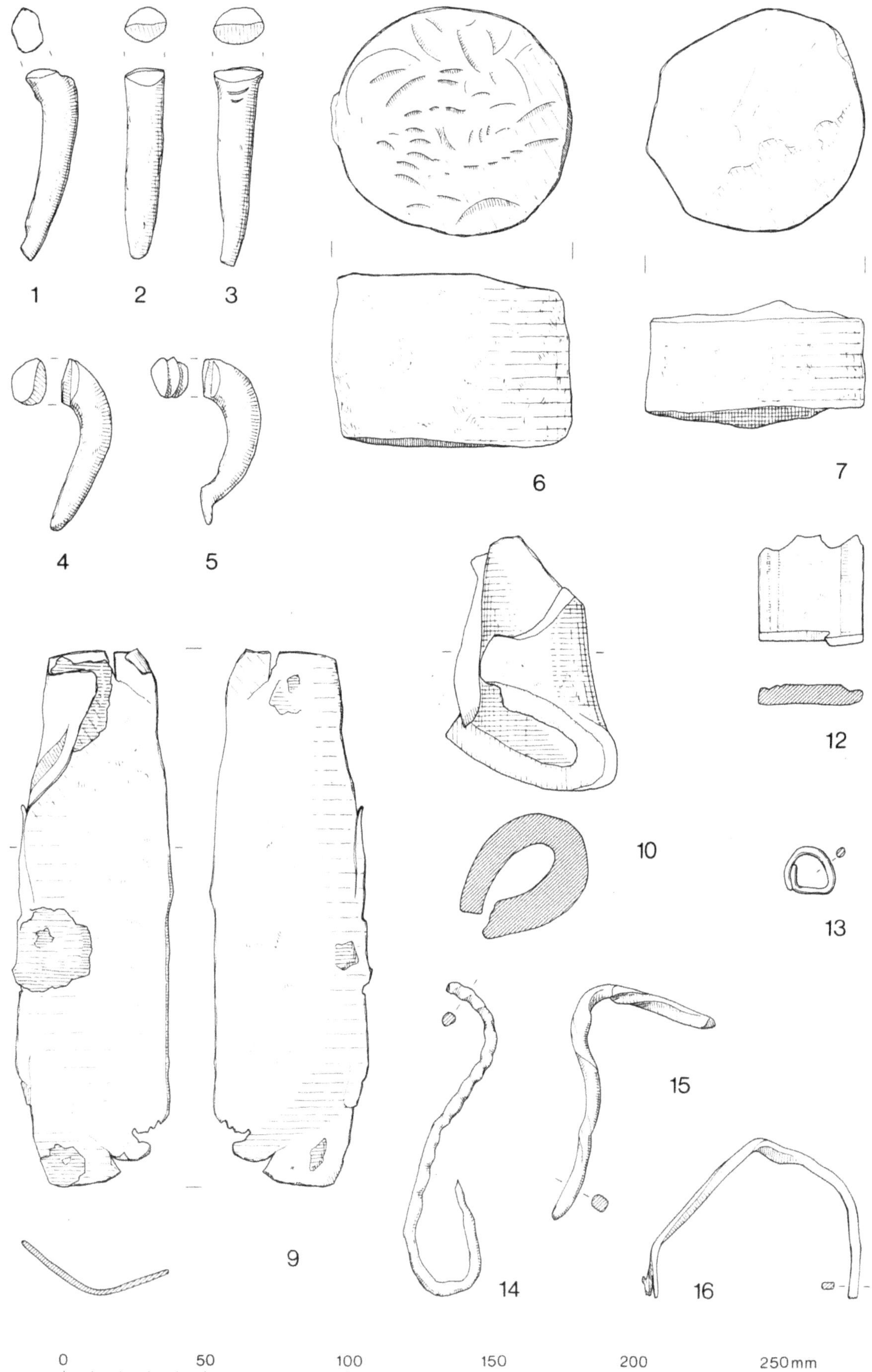

Figure 24: Objects of lead, scale 1:2

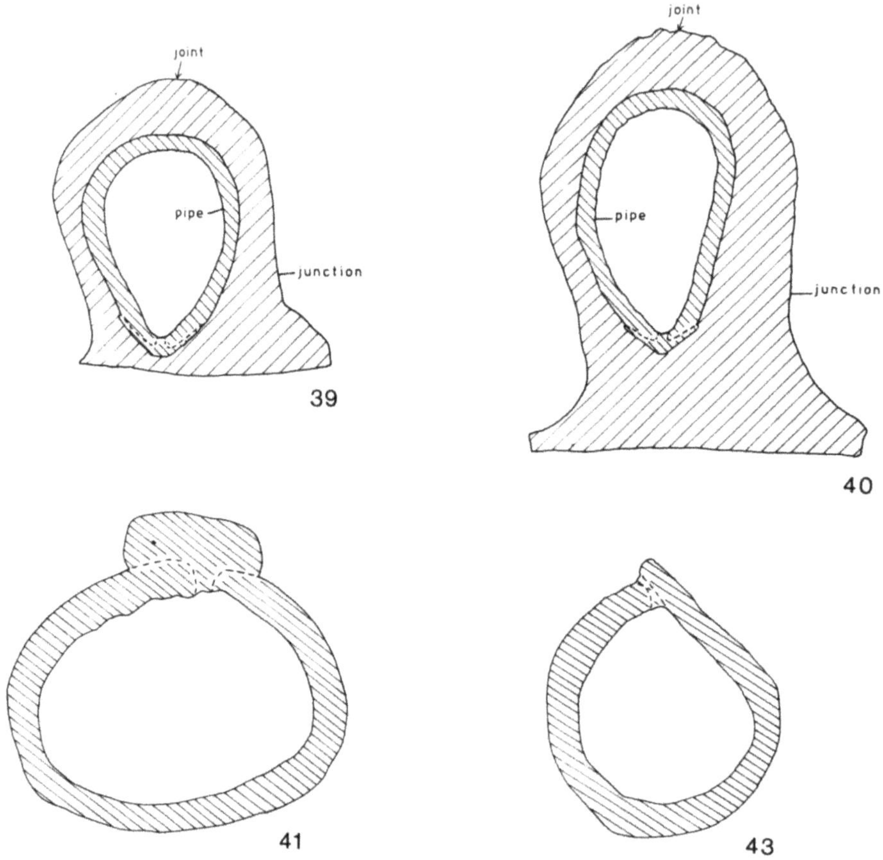

Figure 25: Cross-sections through water-pipes and junctions, scale 1:2

OBJECTS OF IRON (FIGS 26-7)

By Janet Webster and P V Webster (1976-7)

With the exception of nails, the excavations at Pentre Farm yielded relatively little ironwork. The examples found seem to have been connected mainly with the structure or fittings of the building. There is a marked scarcity of domestic equipment or tools which is hard to reconcile with the relative plenty of other domestic rubbish such as pottery.

This section has benefitted greatly from discussion with Professor W H Manning to whom we should like to record our gratitude.

1. Hippo-sandle, missing one wing but otherwise complete. (10/5, ? Stage V). It has a hooked rod (now fractured) at the front, a hooked heel and the one remaining wing bends forward to retain the front of the hoof. The standard classification (by Aubert) is outlined and discussed by Manning (1972, 171, nos 24-5). This example corresponds to Aubert type 1.
2. Large tumbler lock slide key. (0/1, demolition between Stage VII and VIII). The workings of this type of key and its lock are described in Manning (1972, 181) and British Museum (1920, 149-50). A tumbler lock key of similar dimensions is given in Webster (1975, no 221).
3. Tumbler lock lift key. (0/6, demolition between Stages VII and VIII).
4. Tumbler lock lift key. (0/1, demolition between Stages VII and VIII). This simpler form of tumbler lock can be found in Manning (1972), British Museum (1920) and Webster (1975).
5. T-shaped item flattened on the stem of the T and evidently fixed to some larger object with two rivets, (topsoil). It probably served as a leg with a splayed foot; attachment to a grid-iron or brazier seems possible.
6. Foot, probably from a grid-iron, (topsoil). An example, from Great Chesterford, of the complete object is given by Liversidge (1968, fig 65, d).
7. Functional ring. (1/30, Stage VII). Various uses are of course possible.
8. Bar with hammered knob at one end. (6/57, isolated deposit in North-west Range, Stage V). There appears to be a swelling below the knob suggesting that this is a handle of some tool or

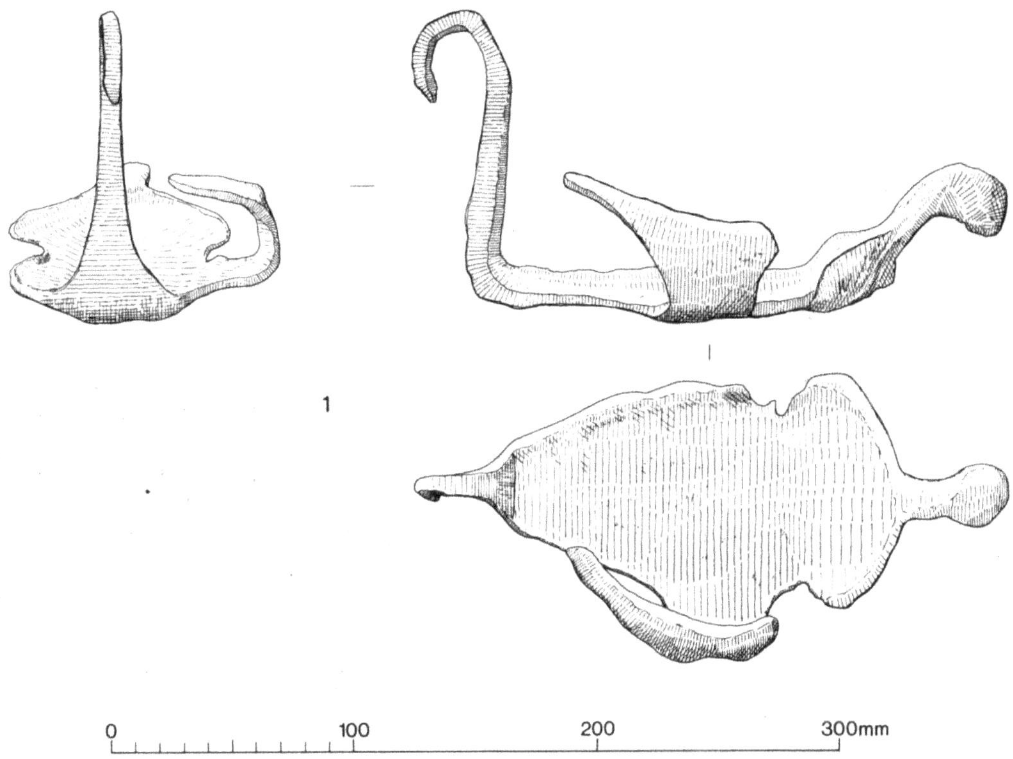

Figure 26: Objects of iron, scale 1:3

implement. An item like a poker or fire shovel seems possible although there is no evidence on the extant fragment of the twisting characteristic of this type of object.

9. Split-spike loop. (0/25, demolition between Stages VII and VIII). The split-spike loop was generally driven through timber and the spikes opened out and hammered over when they had penetrated to the other side; timber of at least 50 mm thickness is thus implied by this example. A discussion of the type with examples is presented by Manning (1972, no 90).
10. Fragment of nailed strip or binding. (2/37, Stage V). A similar piece was found at Gadebridge Park (Manning 1974, no 595).
11. Joiner's dog. (0/1, demolition between Stage VII and VIII); cf Manning (1974, nos 653-4).
12. Part of a T-shaped staple. (2/21, Stage VII). This is a common means of securing box-tiles in position and seems the most likely function in the present context; cf Manning (1974, nos 511-17).
13. Part of a hook. (7/11, isolated deposit in North-west Range, Stage VII). Manning (1974, nos 522-3) gives complete examples of the type from Gadebridge Park.
14. Iron bar. Some structural use seems likely, possibly as a lintel in a position where a wooden lintel would be exposed to fire damage (eg in a furnace flue). (Unstratified).
15. Iron point, heavily corroded and eroded away from the point. (9/17, Stage VII). Part of a larger object (see no 16).
16. Portion of an unidentified object. (1/7, Stage VIII or later). The complete piece probably had two pointed 'legs', one at either end of the cross-bar, the whole being not unlike a modern boot scraper in appearance. Fragments of two other 'legs' imply at least one other similar object. The similarity of the points with that on no 15 is striking. A Roman date seems to be implied by the context of at least some of these objects; structural use, presumably as a giant staple, seems probable.

Nails (Not illustrated) (PVW)

Very few nails were recovered from stratified levels, suggesting perhaps that wooden elements in the earlier phases were pegged together rather than nailed. The only significant quantity of nails comes from the uppermost levels where the nails presumably derive largely from the destruction of the building. All nails examined were heavily corroded, and most were fragmented. It was, however, possible to make an approximate division according to the classification suggested by Manning (1976, 41, fig 11). All identifiable nails fell within Manning's type I. This type can be subdivided according to nail head shape: type Ia has a domed head, while type Ib is flat-headed. All the nails from upper levels at Pentre Farm have been grouped according to this classification (Table 2).

Nail type	Number	%	Weight (kg)	%
Type 1a	19	12	0.42	18
Type 1b	76	49	1.02	44
Corroded etc	60	38	0.86	37
Total	155	99	2.30	99

Table 2: Classification of nails

It cannot be claimed that these figures represent a very large type sample, but they seem internally consistent. Type Ia tends to be larger and therefore heavier than Ib and this is reflected in the figures. The unidentifiable pieces should consist of fragments of both types and thus the percentage difference between weight and number should be, and is, less.

Many of the nails examined show some buckling but very few exhibit the characteristic curve of the nail extracted with a clawed tool. It seems likely on this evidence that the structure of which they formed part was allowed to decay and collapse rather than being deliberately demolished while its timbers were still re-usable.

OBJECTS OF BONE (FIG 28)

By Janet Webster (1976-7)

1. Dice. (SF 416, 10/100, midden outside Rooms 24-25, Stages II-III). Numbered from one to six by means of motifs consisting of an inscribed dot and two rings. The dice is polished smooth. (For examples see Neal 1974, 157, fig 67, no 336; Hassall and Rhodes 1974, 72, fig 28, nos 34-5; Curle 1911, pl XCIII, no 3; Wheeler and Wheeler 1928, fig 14, no 27).
2. Counter with carved concentric circle design. (SF 117, 13/14, yard to the north of the complex, Stage VII). (Curle 1911, pl XCIII, nos 4, 24-5).
3. Counter with dished face, central incised dot and faint traces of incised concentric circle decoration. (SF 123, 0/32, post-complex).
4. Highly polished bone counter with incised dot on one face. (SF 380, topsoil above the complex).
5. The curved handle of an iron implement. (SF 128, 0/1, post-complex). Made of animal horn, broken but with the iron tang remaining. The tip of the horn was cut off and the whole facetted lengthways when it was fashioned into a handle.
6. Needle with oblong eye and point missing. (SF 132, 0/32, post-complex). It has a uniform, though obviously tapering, shaft so that the cross-section at the eye is the same as lower down the shaft. This type also terminates in a stubby point above the eye (Wheeler 1930, pl XLII, nos 7, 9).
7. A more complete example than no 6 above. (SF 253, 3/30, Room 52/3, Stage VIII).
8. A further needle of the type described in no 6 with the point lost. (SF 106, 3/5, 7 Stage VI).
9. (Not illustrated). Bone needle fragment, probably of the same type as no 6. (6SF 328, 6/20, Room 29, Stage IV).
10. Needle with the eye cut in a broad flat head terminating in a rounded tip. (SF 76, 0/1, post-complex). (Wheeler 1930, pl XLII, no 8). In this case the eye consists of two circular piercings, one above the other and running into each other. The point is lost. (A needle with a similar eye is given in Hassall and Rhodes 1974, 72, fig 28, no 14).
11. Another example of the type described in no 10. (SF 107, 0/28, post-complex). The point is missing. The eye is oblong.
12. Pin with simple knobbed head and swelling shaft. (SF 88, 0/3, post-complex).
13. Similar, larger pin, with point intact. (SF 48, topsoil above the complex).
14. Pin with elaborately carved head consisting of a conical knob with two collar mouldings below. (SF 43, topsoil above the complex).
15. Part of a pin with a faceted shaft and crudely carved head consisting of conical knob and collar moulding. (SF 74, topsoil above the complex).
16. Pin with neatly carved head consisting of a conical knob and collar moulding. (SF 376, topsoil above the complex).

17-18. Heavy bone pins, both broken. (SF 60, topsoil above and complex, and SF 116, 0/32, post-complex).

19. Similar to nos 17-8 but not as heavy. (SF 75, 0/1, post-complex).

20-26. (Not illustrated). Seven further bone pin fragments were recovered from various contexts.

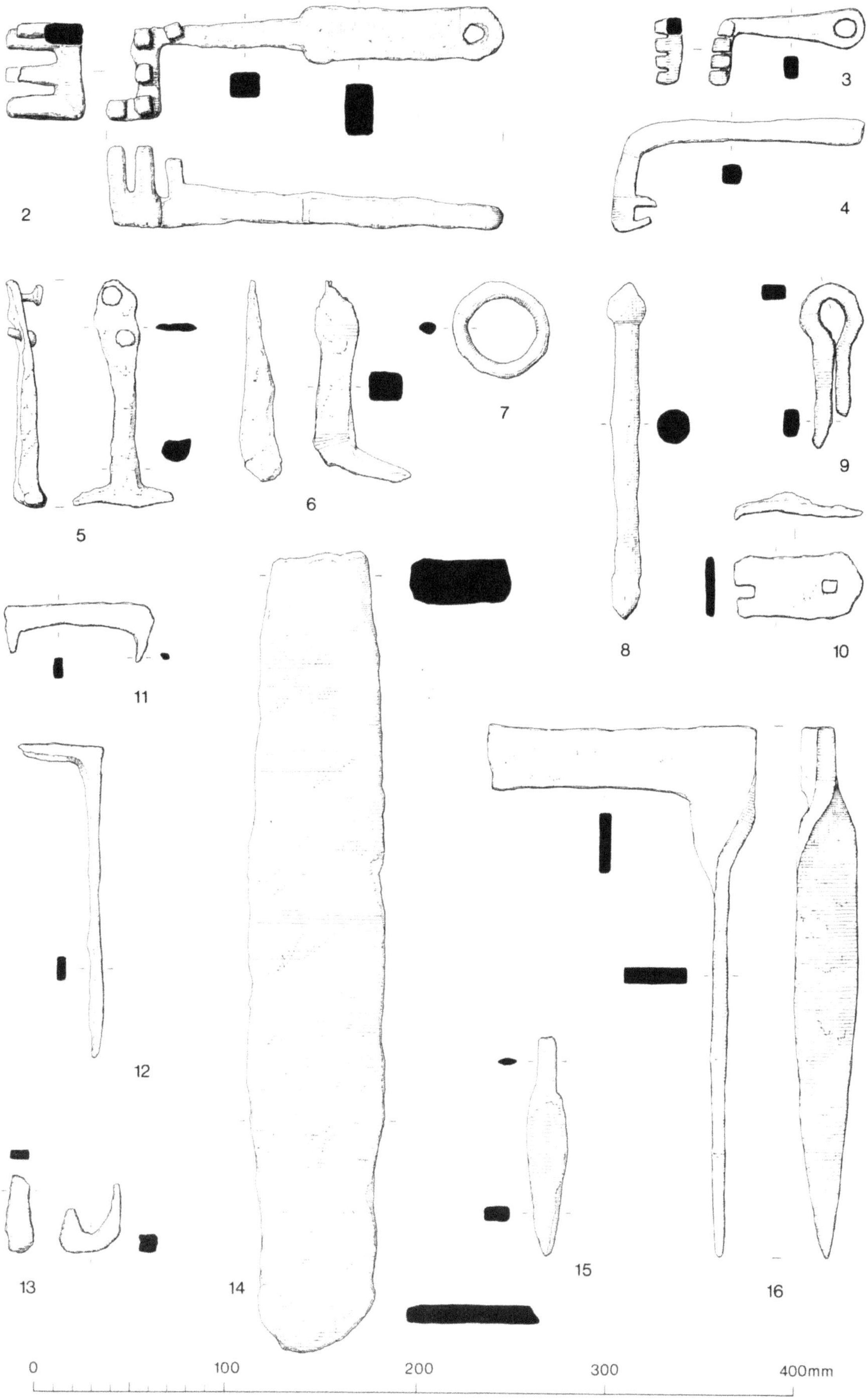

Figure 27: Objects of iron, scale 1:3

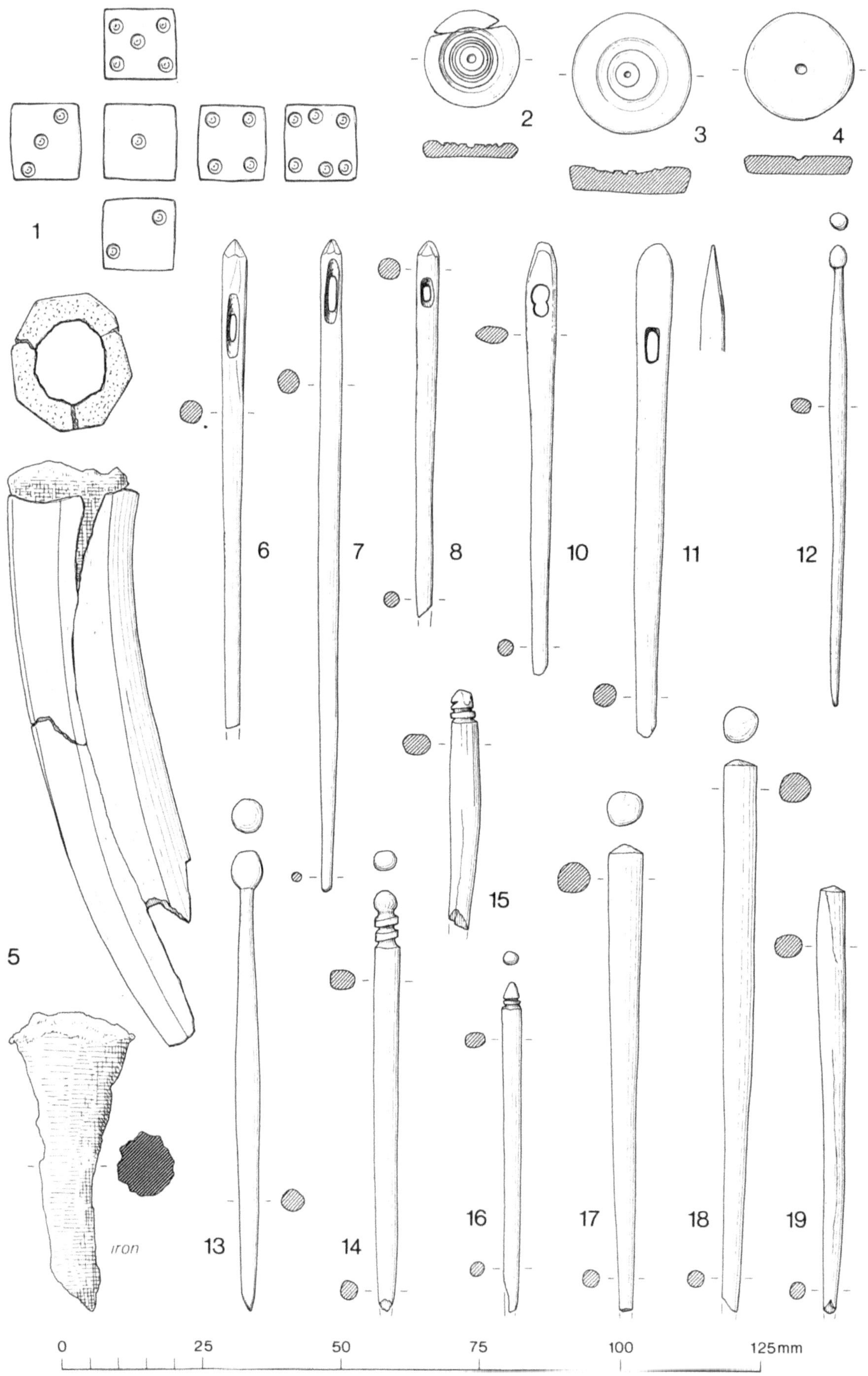

Figure 28: Objects of bone, full size

THE GLASS

By Jennifer Price (1987)

Discussion

During the excavations a total of 301 fragments of Roman glass were found. Of these, 201 were from vessels, eighty-nine were window glass and eleven were objects. The vessel glass was bluish green, except for fifteen coloured and forty-one colourless fragments, as was the window glass, apart from four colourless fragments.

The vessel glass has been divided into two main categories - coloured, colourless or bluish green tablewares, such as drinking cups, bowls, jugs, flasks and bottles (118 fragments), and bluish green household containers (eighty-three fragments). The tablewares are generally of good quality, some of them being quite fine, and there are several rather unusual vessels in this assemblage (notably nos 1, 2, and 24).

Among the tableware fragments a minimum of four coloured, sixteen colourless and five bluish green vessels have been recognised, and there is certainly one cylindrical bottle among the containers, but it has proved more difficult to establish the number of square bottles. There may be as few as four or five, though probably many more are represented by the surviving fragments.

The vessel forms represented range in their dates of production from the later first century at the earliest to the middle of the third century at the latest. In general, strongly coloured glass is found only rarely after the Neronian to early Flavian period, but yellowish brown and yellowish green vessels, usually jugs (cf nos 3-4), sometimes occur in mid second century deposits, and peacock blue vessels, (nos 1-2) have occasionally been noted on second century sites, so the coloured vessels used at Pentre Farm may have been in production during the early to mid second century. All of the colourless vessels could have been in use during the second century, though the facet-cut beaker (no 5) may have been made in the late first century, and the cylindrical cups with fire-rounded rims (nos 12-17), jug with pouring spout (no 21) and cylindrical jug or bottle (no 23), could have been produced in the early third century.

The bluish green tablewares and containers most probably also belong to the second century, though jugs similar to no 25 and bowls similar to nos 26-7 occur in Flavian contexts. Square bottles are very commonly found in large quantities on all kinds of Romano-British sites from the Neronian period to the end of the second century, while cylindrical bottles disappear early in the second century. The decorated cylindrical bottle (no 24) is a most unusual find in Roman Britain. It unfortunately comes from a late Roman context, but there is little doubt that it should be dated to the Flavian-Trajanic period, and so it was probably in use on the site in the early part of the second century AD.

The objects, counters (or gaming pieces) and beads, cannot be dated very closely as they occur throughout the Roman period, though none is necessarily later than the second century AD. The window fragments all come from cast panes; cast window glass was produced in the first and second, and perhaps the early third centuries, before being replaced by blown window glass in the third and fourth centuries.

The only fragment of cast vessel glass comes from a peacock blue bowl with overhanging rim (no 1). Cast vessels in this colour are not very common in Roman Britain, though a shallow convex curved bowl with horizontal ribs on the body and a wheel-cut groove inside the rim is known from Trinity Street, Dorchester, Dorset (Cool & Price forthcoming), and curved body fragments have come from the Lloyds Bank site, Dorchester, Dorset and from the Hermitage, Box, Wiltshire.

Bowls with overhanging rims were usually made in colourless glass. They do not occur before the Flavian period, but are found throughout the Roman world in the late first and early second century. This vessel type is well-known in Roman Britain, especially in Flavian-Hadrianic contexts, as at Fishbourne (Harden & Price 1971, 331 & 334, nos 26, 28, 33-4), and numerous sites in northern Britain (see, for example, Charlesworth 1959, 38-40).

Comparatively few bowls with overhanging rims made in coloured glass have been found in Britain. An opaque red fragment was found at Ditchley, Oxon (Harden 1936, 64, fig 12.3), and a small shallow dark yellow brown plate is known from King William Street, London (not dark blue as stated by Wheeler 1930, 122, fig 42.2). Two peacock blue fragments, from bowl rims without an overhang, have been noted from Welbeck Street, Castleford, West Yorkshire (unpublished) and Birrens (Robertson 1975, 137, no 39, fig 48.3). None of these came from dated contexts, but it seems probable that the coloured specimens were in use at the same time as the colourless ones (that is, the late first and early second century).

It is interesting that another vessel, the fragmentary flask with globular body and funnel mouth (no 2), was also made in peacock blue glass, as blown vessels in this colour are not at all common in Roman Britain, though there are some fragments from a shallow bowl at Cosgrove, Northants (Price forthcoming). The form of the vessel is also rather unusual; its colour, rim finish and decoration suggest strongly that it is a first or early second century vessel, but it has not been possible to find any close parallels.

Most of the globular flasks with funnel mouths found in the north-western provinces are larger than this specimen. One variant with a slender long neck and a small funnel mouth is known in Hadrianic to late second century contexts (Isings 1957, Form 92), and another, which sometimes has a base ring, belongs to the third and fourth centuries (Isings 1957, Form 104;

Harden & Green 1978, 163-170).

A colourless flask from Cologne (Harden *et al* 1968, no 114 - described as a third century indented beaker) is closely comparable with the Pentre Farm example. The rims, mouths, necks and the form and wheel-cut decoration of the shoulders of the two vessels are very similar, and these features are characteristic of colourless drinking cup forms made in the late first and second century. The rest of the Cologne flask, however, is quite different, with raised arcading and six indents on the ovoid body and a small tubular base ring with domed base. Despite the different coloured glass, the two vessels appear to belong to the same tradition of manufacture, and both could have been produced at about the same time, around the end of the first or in the early to mid second century.

Nos 3 and 4 are likely to come from conical long-necked jugs, though the small size of the fragments makes the identification uncertain, and no 25 very probably also belongs to a vessel of this form (Isings 1957 Form 55). These jugs, which have a small folded rim, long narrow neck, conical body and an angular ribbon with central rib, usually with a vertical trail extending down the body from the handle, and either a simple concave base or an open base-ring, occur in the region north of the Alps but not elsewhere in the Roman world. A number of broadly contemporary jug forms were made in bright colours such as dark blue, yellowish brown, yellowish green and purple, as well as bluish green glass. These are frequently found in contexts from *c* AD 65 to the early second century, though some examples, including a few yellowish green and yellowish brown jugs, remained in use until at least the middle of the second century (Price 1980, 66; Price 1987, 193-5).

The form is very common on Romano-British sites; many examples with the concave base are known, as at Barnwell, Cambs (Harden *et al* 1968, no 73), Radnage, Bucks (Harden *et al* 1987, no 68) and Winchester (Harden 1967), and specimens with open base-rings also occur frequently, as at Towcester, Northants (Price 1980, fig 15, nos 7-9), and Bartlow Hills, Essex (Gage 1834, pl 2, fig 1).

The fragmentary facet-cut beaker (no 5) is a low wide truncated conical vessel with moulding at the rim, pronounced ridges above and below the decoration zone which contains three rows of large oval facets in quincunx, closely set to form diamonds, and a projecting foot. This is a recognised variant within a larger group of similarly decorated high-quality cups which have been found in many parts of the Roman world in the later first century and first half of the second century (Berger 1960, Form IV, 67-70; Welker 1974, 55-62, nos 147-8; Oliver 1984, no 47, fig 5, no 51).

One intact cup, from Barnwell, Cambs (Harden *et al* 1968, no 101; Harden *et al* 1987, no 104) and many fragments have been found in Flavian to Hadrianic contexts in Britain, though comparatively few can be shown to come from low vessels similar to this example. Several pieces have been found at Welbeck Street, Castleford, West Yorkshire (unpublished), and a lower body fragment from Eastgate, Gloucester probably comes from another example (Price 1983, 169, fig 98.11).

Nos 6-10 are pieces from colourless drinking cups with wheel-cut or abraded linear decoration. A very large number of similar vessels have been found in Britain, generally in early to mid second century contexts, but until recently they have attracted little attention, probably because they usually occur as very small fragments, and the complete profile can rarely be reconstructed (for example, see Charlesworth 1959, fig 7, nos 6, 10-11, pl 11,2; Price 1980, 63-4, fig 14, nos 3-5, & 68, fig 17, no 3; Price 1987, 188-191, fig 2, nos 8-14).

The wheel-cut decoration is sometimes formed by grinding away the outside surface to leave horizontal ribs in relief, as on nos 6-7, but this would probably have been reserved for vessels of considerable luxury, as it must have been a time-consuming process. It has not often been noted on colourless blown cups, though a similar treatment occurs on a shallow segmental bowl from an Antonine pit at Felmongers, Harlow, Essex (Price 1987, 187-188, fig 1, no 1).

No 8, a rim fragment, is closely comparable with a globular cup from a second century burial at Braughing, Herts (Harden 1977, 102, fig 43.3, pl IXB, and fragments from a deposit dated *c* AD 160 at Shakenoak, Oxon (Harden 1971, 101, fig 44.45-6, nos 98-9). The base fragments, no 9, also come from a wheel-cut drinking cup; similar bases are known from Fishbourne (Harden and Price 1971, 346, fig 140.59), and Castlecary (which has a wheel-cut ring on the base, though it is not shown in Charlesworth 1959, fig 7.6). The form of the complete vessel is not known in the case of no 10, a body fragment with twelve bands of abraded lines, but it may have been similar to a piece with multiple close-set abraded bands from the second century deposit in the latrine drain of the Commandant's house at Housesteads (Charlesworth 1971, 35, fig 8). In 1971, Charlesworth stated that this drain deposit was closely dated between AD 128 and *c* 139-42, but a later publication (Charlesworth 1975, 24) mentions later second century coins and pottery from the same drain, so this not a very closely dated group.

The rim fragment, no 11, is from a rather different kind of cup which probably also belongs to the second century, though the type may have continued in production later than the ones discussed above. It is cylindrical with a simple vertical rim which has been ground very smooth, and should be compared with similar rim fragments from Maryport, Cumbria (Price 1976, 52, fig 9.5) and Rudston, North Humberside (Charlesworth 1980, 125, fig 84.2).

Nos 12-17 come from cylindrical cups with fire rounded rims which occur in large quantities in the north-west provinces during the later second and early third centuries (Isings 1957, Form 85). Nos 12-17 are from the undecorated form with a vertical or slightly inturned rim and a tubular and trailed double base ring, which is the commonest type of drinking vessel in use in Roman Britain during this period. One example has been found intact at Airlie, Angus (Charlesworth 1959, pl I.4) and fragments are known on many other sites, including Fishbourne (Harden & Price 1971, 352-5, fig 141, 74-5), the fortress baths at Caerleon (Allen 1986, 111, fig 43, nos 68-70,), York (Harden 1962, 137, fig 86.HG 202.6), and Housesteads (Charlesworth 1971, 34, figs 4 & 5 - but see the discussion of the dating of this deposit in connection with no 10, above).

The other version (no 12), which has a slightly outbent rim, horizontal trails on the upper and lower body and a variety of base rings, including the double base ring, is also very widely known. A complete cup was found at Baldock, Herts (Charlesworth 1959, 49, pl II.1) and fragments have come from a wide range of sites, as at Caerwent (Boon 1972-3, 121, fig 4.41), and in the second century drain deposit at the Commandant's house, Housesteads (Charlesworth 1971, figs 1-2, and the base only of fig 3).

The colourless tubular rim (no 18) and the bluish green tubular rim (no 27) very probably come from the same kind of bowl, with a more or less cylindrical body and an applied 'true' base ring (cf no 26). Bowls with tubular rims were produced over a long period of time (Isings 1957, Forms 44-5). The earliest examples in Britain date from the Claudian period at Camulodunum (Harden 1947, pl LXXXVIII.80-1) and the latest occur in the later fourth/early fifth century deposit at Burgh Castle (Harden 1983, fig 37, 81-2), but most of these bowls are found during the later first and second century AD, as at Richborough, Kent (Bushe-Fox 1949, pl LXVIII.369 & 372), and Long Melford, Suffolk (Avent & Howlett 1980, 246, fig 41, pl XVIIIa). These bowls were produced in coloured and bluish green glass, and were sometimes decorated with ribs on the body. Colourless examples are quite uncommon, though a pale greenish colourless bowl thought to date from the end of the first century is known at Ell-Hunsel, Limburg, Netherlands (Isings 1971, 20, fig 4, no 55).

The rim fragment, no 19, and base fragment, no 20, come from colourless small bowls, perhaps like two from Hauxton, Cambs (Harden 1958, 12-3, 15, fig 7, pl III, c-d). Small bowls of this kind, sometimes with a ridge on the underside of the rim, have been found in second century contexts at Maastricht-Belfort and elsewhere in Limburg, Netherlands (Isings 1971, 21, fig 3, no 57, 23, fig 12, no 66, 8).

The small rim fragment, no 21, very probably comes from a jug with a trefoil mouth or pouring spout, though it is too small for certain identification of the form. Jugs with trefoil mouths occur at many periods in the Roman empire (cf Isings 1957, Forms 56 and 88), but in Britain they are most frequently noted in second to early third century contexts, when they may have folded or rounded rims. A greenish jug with folded rim similar to the Pentre Farm fragment is known from a second century burial at Skeleton Green, Herts (Charlesworth 1981, 271, fig 106.12), and others with fire-rounded rims include examples from Ospringe (Whiting *et al* 1931, pl XVII.154) and the fortress baths at Caerleon (Allen 1986, 108-9, fig 42.57).

The tall narrow neck, no 22, may come from a variety of flask forms, so nothing can be said about the precise type of vessel. Similar fragments are known from many sites. The neck may be compared with that of the ovoid bodied flask from Hauxton (Harden 1958, 12-14, fig 5, pl IIIa).

The body fragment, no 23, probably comes from a colourless cylindrical bottle with horizontal bands of linear cut decoration on the body. These occur in second and third century contexts in the north-west provinces, and they have been noted on several sites in Britain.

A complete one is known from Hauxton (Harden 1958, 12, 14-5, fig 6, pl IIIb), and there is a large fragment from another at Corbridge (Charlesworth 1959, 54, fig 10.1). The vertical scratching visible on the shoulder angle suggests that the vessel was stored in a close-fitting case.

The fragmentary bluish green cylindrical bottle with horizontal bands of linear wheel cut decoration, no 24, is a most unusual find on a Romano-British site. The shape, details of manufacture and colour of the vessel are precisely similar to the undecorated cylindrical bottles discussed below as containers, but I am not aware of any other examples with wheel cut decoration in the north-western provinces. Isings (1957, Form 51, 67) refers to some specimens at Pompeii which have a decoration of wheel cut lines, but says 'in the provinces this is very rare in early bottles'. They are, however, commonly found in parts of north Africa and the eastern Mediterranean area, and it is possible that the vessel travelled a long way to North Wales, probably as a personal possession, rather than as an object of trade. The vertical scratches on the body and heavy wear on the base edge suggest that it was stored in a close-fitting case and that it was in use for a considerable time.

The rim fragment, no 28, probably comes from a small bowl or jar, but it is not easy to identify the exact form from the surviving piece. A considerable variety of jars are found in second century contexts, but they have not received a great deal of attention in print. They are very rarely found complete, unless they occur as grave goods or have been used as cinerary urns (cf Isings 1957, Form 94).

The fragments of containers (nos 29-33) come from bluish green cylindrical or square bottles. Both forms have similar folded rims, necks, shoulders and broad angular handles, often with multiple reeding on the lower part. The bodies often have vertical scratch marks, suggesting that they have been moved in and out of a close-fitting case, and the necks sometimes have a band of horizontal scratch marks, probably indicating that the stopper was attached to the bottle by a cord. Both forms are extremely common finds on Romano-British sites, and occur throughout the western provinces in the first and second centuries AD (Isings 1957, Forms 50-51).

The cylindrical bottles occur very widely on first century sites, particularly in Flavian contexts, and they are not common after the first quarter of the second century. Two main forms are known; one has a tall and comparatively narrow body with a short narrow neck, as at Caerleon (Boon 1967, 95, fig 1) and Inchtuthil (Price 1985, 307, fig 94.11), and the other has a low wide body, often with a wide neck, as at York (Harden 1962, pl 66.HG 53). No 31 cannot be reconstructed from the surviving fragments, but the body diameter suggests that this was a tall bottle.

It is noteworthy that only one cylindrical bottle has been found, which indicates that square bottles were the main containers in use during the occupation of Pentre Farm, and it is also interesting that one of the body fragments shows clear evidence for re-use, as the broken surface is completely flat and matt with a great deal of wear. The re-use of the edges of fragments of thick glass has not often been noted, though two edges from a piece of a large square bottle at Cosgrove had been re-worked in this way, and it was there suggested that the piece may have been used as a palette, or for some industrial purpose, such as the preparation of hides (Price forthcoming).

Square bottles are also very commonly found on first century sites, but, unlike the cylindrical bottles, they continue in widespread use until at least the end of the second century (Charlesworth 1966). Square bottles were usually blown into a body mould with a separate decorated base mould, and a very wide variety of raised designs occur on the bases (see Charlesworth 1959, fig 9). Most of these bottles are well-made and of robust construction, as they would need to be for the long-distance transport of liquid and semi-liquid substances, but no 33, with three concentric circles in low relief, comes from a small bubbly bottle which has been blown very thin at the angles. These small, carelessly made and rather fragile bottles have been noted at several first and second century sites in Roman Britain, and may have been the products of local glassmaking centres, as they were presumably used as containers for the transport of only very small quantities of substances and they were unlikely to have travelled more than quite short distances. A group of these small carelessly made square bottles has been recognised from Maidstone, Staines, Colchester and Harlow (Price 1987, 197-199, fig 4, nos 33-34), and further work on the base designs will identify other groups.

The objects fall into two categories: counters or gaming pieces, and beads. Two of the counters (nos 35-6) are the usual 'black' and opaque white plano-convex discs which occur on British sites throughout the Roman period, generally as single finds, though sets for board games are sometimes found in burials, as in a Flavian burial at Grange Road, Winchester (Biddle 1967, 244-5, fig 9) and a late second to early third century burial at Ospringe (Whiting *et al* 1931, pl LVI). The third, no 34, is a much less common type: it is opaque turquoise blue, sub-rectangular in shape and completely smooth on the base, and there is a wavy band at the edge of the convex surface which may be the marks of a setting, perhaps from a ring or pendant.

The glass melon bead, no 37, is a type of bead much more widely known in frit or faience which occurs throughout the Roman period. The glass versions occur in some quantities during the later first and early to mid second century (Guido 1978, 100, fig 37.21-2), and are often found on military sites. The biconical bead, no 38, (Guido 1978, 97, fig 37.12-3) is a type frequently found in late Roman contexts, but also known at earlier periods. Four were found in deposits dated to *c* AD 160-230 in the drain of the legionary bath-house at Caerleon (Brewer 1986, 148, fig 48.62-5). The long biconical bead, no 39, (Guido 1978, 98, fig 37.14), and the pentagonal beads, nos 40-3, (Guido 1978 96-7, fig 37.8) occur throughout the Roman period, so they cannot be closely dated.

The window glass is of the cast, matt-glossy, type found at very many sites in Britain in the first to early third century (Boon 1966). The pieces at Pentre Farm are made in various shades of bluish green and near colourless glass, and are often uneven in thickness, being thin in the middle and much thicker near to the edges. They are flat, dull and pock-marked underneath and uneven and bright on the top. Six pieces have a thick rounded edge, sometimes with tooling marks in the top surface, and two joined fragments have a grosed edge, similar to the windowpane from Hartfield, Sussex (Harden 1974). Some of the edge fragments have traces of the mortar, which fixed them into the window aperture or wooden frame, on the underside. As only very little of this material comes from stratified contexts, it is not possible to identify many of the rooms with glazed windows.

Catalogue (Figs 29-31)

A more detailed catalogue of the fragments not illustrated is housed in the site archive.

Cast and Ground Vessels

1. Rim fragment, bowl with overhanging rim. Deep greenish blue (peacock blue). Very few small round bubbles; little visible weathering, except for slightly pitted surfaces. Part of broad diagonally outsplayed rim with overhanging edge, straight side tapering inwards. Upper rim surface ground flat, leaving ridges at junctions with overhang

and body. All surfaces ground and rotary-polished. Present height 11 mm; rim diameter 160 mm, thickness 1-2.5 mm. SF 348, 6/51, Stage IV, Room 29 (late 2nd/early 3rd AD).

Blown vessels - coloured

2. Ten rim and body fragments, seven joined, flask with globular body. Deep greenish blue (peacock blue). Small round bubbles, slightly elongated in neck; flaking iridescent weathering, pitted with small fissures. Curved rim, edge cracked off and ground, funnel mouth, short neck, wide convex curved body and part of lower body tapering towards base (missing). Narrow wheel cut line below shoulder, two bands of wheel cut and abraded lines on the upper body, two close-set narrow wheel cut lines on lower body. Present height 72 mm; rim diameter 49 mm; thickness 0.5-2 mm. SF 413, 7/88, post-complex.
3. Two joined body fragments, probably from a conical jug. Light greenish yellow. Small round bubbles; little visible weathering, except for slightly rough patches on outside surface. Part of straight side with one broad vertical rib. Present height 59 mm, thickness 1-1.5 mm. SF 325, 80, post-complex.
4. (Not illustrated). Two small body fragments. Very pale greenish yellow. Small elongated bubbles; no visible weathering. Part of straight side with faint evidence for one rib. Dimensions 11 x 13 mm; thickness 1.5 mm. SF 391, 6/13, post-complex.

Blown vessels - colourless: with wheel cut decoration

5. Six rim, body and base fragments, four joined, conical beaker. Colourless. Very few round small bubbles; lightly pitted outside surfaces, strain crack. Rim edge ground and polished, straight side tapering in to foot, base missing. Outside surface cut away, leaving two shallow horizontal mouldings at the rim and one on the upper body, a decoration zone, and a deep horizontal ridge on lower body above the diagonally projecting foot. The raised decoration zone has three rows of large oval facets cut in quincunx, close-set to form diamonds; the top row has small oval facets between the large ones. Vertical polishing marks are visible within the facets. Height 83 mm; rim diameter 92 mm; base diameter 54 mm; thickness 2-8 mm. SF 410b, 2/53 & SF 417c, 10/100, post-complex.
6. Rim fragment, hemispherical cup. Colourless. Very few small round bubbles; dull, lightly pitted outside surface, strain cracks. Small curved rim, edge cracked off and ground smooth, convex curved body. Two horizontal wheel cut lines at rim; outside surface wheel polished. Present height 35 mm; rim diameter 86 mm; thickness 1-1.5 mm. SF 375, 10/30
7. Rim fragment, cup. Colourless. Very few small round bubbles; bright, lightly pitted outside surface, strain crack. Small curved rim, edge cracked off and ground smooth, slightly convex curved body. Horizontal wheel cut line and raised rib below rim; outside surface wheel polished. Present height 31 mm; rim diameter 110 mm; thickness 1 mm. SF 412, 5/21.

Fragments with wheel polished outside surfaces. (Not illustrated). Two body fragments, post-complex.

8. Rim fragment, cup. Colourless. Some small bubbles; dull, grey-brown weathering, pitted. Small curved rim, edge cracked off and ground smooth, constriction above convex curved upper body expanding out. Horizontal wheel cut line at constriction. Present height 17 mm; rim diameter 60 mm; thickness 1.5-2 mm. SF 35, 0/64, post-Roman.
9. Three joined fragments, lower body and base, cup. Colourless. Small bubbles; dull, grey-brown weathering, pitted. Small part of cylindrical lower body, tapering in to a flat base. Horizontal wheel cut line on lower body, wheel cut line at edge of base. Present height 13 mm; base diameter 84 mm; thickness 1-2 mm. SF 244, 0/42, post-complex.
10. (Not illustrated) Body fragment. Colourless. Small bubbles; dull, strain crack. Part of slightly convex curved body. Horizontal wheel cut line above twelve narrow diagonal abraded bands. Present height 34 mm; thickness 1-1.5 mm. SF 166, 0/35, post-complex.
11. Rim fragment, cylindrical cup. Colourless. Small bubbles; dull, patches of light pitting. Vertical rim edge, cracked off and ground smooth, straight side. One narrow horizontal abraded band below rim, one on upper body. Present height 27 mm; rim diameter 80 mm; thickness 1-1.5 mm. SF 240, 2/12, post-complex.

Blown vessels - colourless: with trailed decoration

12. Body fragment, cylindrical cup. Colourless. Bubbly; dull, strain cracks. Part of slightly convex curved side. Horizontal trail below rim (missing). Present height 45 mm; thickness 1-1.5 m. SF 347, 7/74, pre Stage IV earlyto mid or late second century AD.

Blown vessels - colourless: undecorated

13. Rim fragment, cylindrical cup. Colourless. Bubbly; pitted. Vertical rim, edge fire rounded and thickened, straight side. Present height 26 mm; rim diameter 84 mm; thickness 1-3 mm. SF 96, 2/1, post Stage VIII - mid third century onwards.

(Not illustrated). Two fragments of rim, as no 13, post-complex.

14. Rim fragment, cylindrical cup. Colourless. Small bubbles, some oval, in horizontal alignment; some scratches. Vertical rim, edge fire rounded and thickened, straight side. Present height 30 mm; rim diameter 90 mm; thickness 2-4 mm. SF 93, 0/12, post-complex.

(Not illustrated). Rim fragment as no 14, post-complex.

15. Rim fragment, cylindrical cup. Colourless. Small bubbles; dull, some pitting. Vertical rim, edge fire rounded and thickened, straight side. Present height 40 mm; rim diameter 96 mm; thickness 1-4 mm. SF 249, topsoil.
16. Base fragment, cylindrical cup. Colourless. Small bubbles; dull. Small part of lower body, tubular outer base ring, trailed inner base ring. Present height 4 mm; base diameter 38 mm; thickness 1+ mm. SF 53, topsoil.
17. Base fragment, cylindrical cup. Greenish colourless. Many small bubbles; dull, some strain cracks. Part of wide lower body, tubular outer base ring, trailed inner base ring, central applied dot. Pontil scar on inner ring. Present height 8 mm; base diameter 44 mm; thickness 1+ mm. SF 11, 0/60, post-Roman.

Body fragments from colourless cups

(Not illustrated). One fragment from Stage V, Room 39, seven fragments post-complex.

18. Rim fragment, bowl with tubular rim. Colourless. Bubbly, with oval bubbles in horizontal alignment; dull, strain cracks. Vertical rim, edge bent out and down to form hollow tube, small part of upper body tapering inwards. Broken edge of body grosed. Present height 10 mm; rim diameter *c* 180 mm; thickness 1-3 mm. SF 424, 9/40, post-complex.
19. Rim fragment, bowl. Colourless. Some small bubbles; dull. Diagonally everted rim, edge fire rounded and thickened, small part of vertical upper body. Present height 12 mm; rim diameter 100 mm; thickness 1-2.5 mm. SF 40, topsoil.
20. Two base fragments, bowl. Colourless. Bubbly; dull. Outsplayed tubular base ring, slightly convex base. Present height 5 mm; base diameter 70 mm; thickness 0.5-2.5 mm. SF 249, topsoil.

Base fragments

(Not illustrated). Two joining fragments came from Stage VII, Room 51, and one fragment from a post-complex context.

21. (Not illustrated). Rim fragment, jug with pouring spout. Colourless. Small bubbles; dull in patches. Very small part of everted rim, edge rolled up and in and flattened, and funnel mouth. Present height *c* 22 mm; thickness 2 mm. SF 7, 0/18, post-Roman.
22. Two joined neck fragments, flask or jug. Colourless. Very bubbly; dull, pitted. Part of tall narrow cylindrical neck with slight constriction at base. Present height 50 mm; neck diameter 20 mm; thickness 2 mm. SF 50, topsoil.
23. (Not illustrated). Body fragment, cylindrical bottle. Colourless. Some bubbles; dull. Small part of curved shoulder and vertical upper body. Vertical usage scratches on body. Dimensions 20 x 31 mm; thickness 1 mm. SF 132, 0/25, post-complex.

Blown vessels - Bluish green: decorated

24. Thirty-nine neck, shoulder, body, base and handle fragments (many joining), cylindrical bottle. Some bubbles, mostly small and round; dull. Very small part of short wide neck, tooling marks at junction with broad convex curved shoulder; wide, almost straight sides, widest at shoulder, concave base (mostly missing). Broad angular ribbon handle with at least seven fine ribs. Five horizontal bands of wheel cut decoration on body; two close-set narrow abraded lines below shoulder, in the middle of the body and above the base, and one broad shallow wheel cut on the upper and lower body. Vertical scratches, especially on shoulder and above base, heavy ring of wear on base. Present height 170 mm (to base of neck); height of body 145 mm; neck diameter *c* 76 mm; maximum body diameter 166 mm (at 140 mm); base diameter 140 mm; thickness 2-4 mm. SF 410a, 2/53, 417b, 10/100 & 410c, IV 2/53, post-complex.
25. Two joined neck and body fragments, conical jug. Some small bubbles, round in body, elongated in neck; no visible weathering except crazing in vertical band on neck. Part of long narrow cylindrical neck, constriction at junction with straight side, expanding out. Close-set ribs, vertical on neck,spiral on body. Present height 108 mm; neck diameter 17 mm; thickness 2.5-5 mm. SF 327, 0/80, & SF 28, 0/60, post-complex and post-Roman.
26. Base fragment, bowl. Occasional small round bubbles; slightly dull. Part of wide convex curved lower body, applied 'true' base-ring, rather distorted. One short diagonal rib above base ring. Present height 15 mm; base diameter *c* 80 mm; thickness 1-4.5 mm. SF 378, 10/42, pre-Stage IV/early to mid or late second century AD.

(Not illustrated). 2 curved body fragments, post-complex.

Blown vessels: undecorated

27. Rim fragment, bowl with tubular rim. Some small round bubbles; little visible weathering. Vertical rim, edge bent out and down to form a round tube, straight side. Present height 15 mm; rim diameter 126 mm; thickness 1-2 mm. SF 323, 0/80, post-complex.
28. Rim fragment, bowl or jar. Some small round

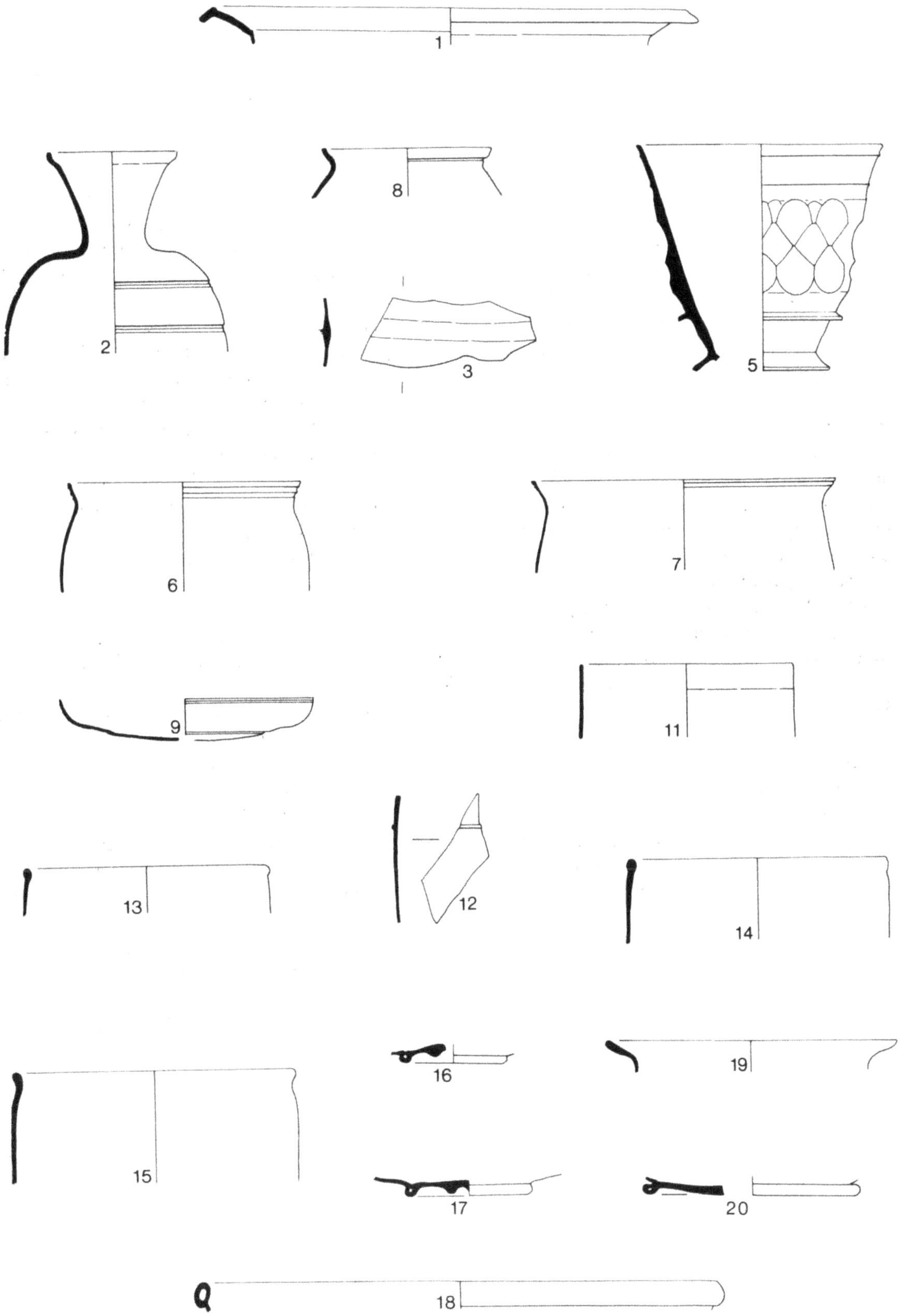

Figure 29: Objects of glass, scale 1:2

bubbles; dull, slight pitting. Wide outsplayed rim, edge bent out and down to form a round tube, slight evidence for change of angle. Present height 10 mm; rim diameter 140 mm; thickness 1-1.5 mm. SF 322, 9/35, post-complex.

Bluish green fragments

(Not illustrated). Three body fragments from post-Roman contexts, six fragments from post-complex contexts, two fragments from topsoil, and two from a Stage VII context.

Blown vessels: bluish green bottles

29. (Not illustrated) Rim and neck fragment. Some bubbles; dull. Diagonal folded rim, edge bent out, up and in, and flattened on the upper surface the edge of cylindrical neck. Present height 23 mm; rim diameter 50 mm; thickness 4.5-5 mm. SF 64, topsoil.
30. Rim and neck fragment. Some bubbles; dull. Horizontal folded rim, edge bent out, up and in, and flattened on upper surface, wide cylindrical neck. Wear scatches on edge of rim and neck. Present height 48 mm; rim diameter 82 mm; thickness 6 mm. SF 291, 3/58, Stage VIII, late second/early third century AD.

(Not illustrated). One fragment from topsoil, one piece from a post-Roman level, one fragment from a context not related to the complex, one piece from post-complex and a fragment from Stage V.

Blown vessels: cylindrical bottles

31. (Not illustrated). Three body and one base fragments. Small round bubbles; dull. Part of shoulder, straight side and concave base. Vertical scratches on body, ring of wear at base edge. NB: one fragment with a very worn and flattened edge has been used for some secondary purpose. Present height (largest fr) 75 mm; body diameter *c* 210 mm; thickness 4-6 mm. SF 325, 0/80, post-complex.

Blown vessels: square bottles

32. Handle fragment. Bubbles and black specks in handle, few bubbles in body; little visible weathering. Part of curved shoulder and flat vertical side; lower part of broad angular ribbon handle with eleven ribs. Present height 51 mm; maximum width of handle 55 m; thickness 1.5-2 mm. SF 299, 8/29, Stage VII - late second to early third AD.
33. Base fragment, small bottle. Bubbly; little visible weathering. Part of slightly concave base with three concentric circles in low relief. Glass very thin near edge of base. Dimensions 28 x 21 mm; diam of outside circle *c* 50 mm; thickness 0.25-2.5 mm. SF 411, 0/6, Room 8, post-complex.

Square bottle fragments

(Not illustrated). Seven fragments post-Roman contexts, forty-nine pieces post-complex levels, nine pieces from topsoil, one fragment from a context not related to the complex, one piece from Stage IV, two pieces from Stage V, one piece from Stage VII, Room 40 and one piece from Stage VIII.

Objects: counters or gaming pieces

34. Sub-rectangular plano-convex disc. Opaque turquoise blue. Wavy band marking edge of convex surface. Flat surface completely smooth, evidence of wear on convex surfaces. Height 6 mm; maximum dimensions 12 x 12 mm. SF 13, 0/60, post-Roman.
35. Plano-convex disc. 'Black'. Bubbly; some pitting. Chip missing from side. Flat surface completely smooth, convex surface dull and worn. Height 6 mm; maximum diameter 15 mm. SF 333, 6/42, Stage IV, Room 30, late second-early third century AD.
36. Plano-convex disc. Opaque white. Bubbly; some pitting. Flat surface dimpled, convex surface dull. Height 7 mm; maximum diameter 22 mm. SF 12, 0/60, post-Roman.

Beads

37. Melon bead. Dark blue. Bubbly; dull. Length 13.5 mm; maximum diameter 18.5 mm. SF 94, 0/20, post-complex.
38. Small biconical bead. Opaque mid blue. Small perforation, two short grooves extending from perforation at one end. Length 6 mm; maximum diameter 8 mm. SF 314, 4/7, Stages VII-VIII, late second/mid third century AD.
39. Long biconical bead. Dark green. Small perforation. Length 7 mm; maximum diameter 4 mm. SF 269, 9/15, Stage V, Room 39, late second/early third century AD.
40. Small five-sided bead. Deep greenish blue (peacock blue). Length 9 mm; maximum width 5 mm. SF 66, topsoil.
41. Small wide five-sided bead. Deep greenish blue (peacock blue). Length 10 mm; maximum width 7 mm. SF 290, CD/52, Stage V, late second/ early third century AD.
42. Two joined fragments, five-sided bead. Dark green over opaque yellow (appearing dark green). Length 12 mm; maximum width 5 mm. SF 72, topsoil.
43. Wide five-sided bead. Dark green. Length 13 mm; maximum width 6 mm. SF 352, 9/40, post-complex.

Window Glass

The window glass is bluish green or nearly colourless, matt on one surface and glossy on the other surface and uneven in thickness. The edges were usually rounded, though two joined pieces have a grosed edge.

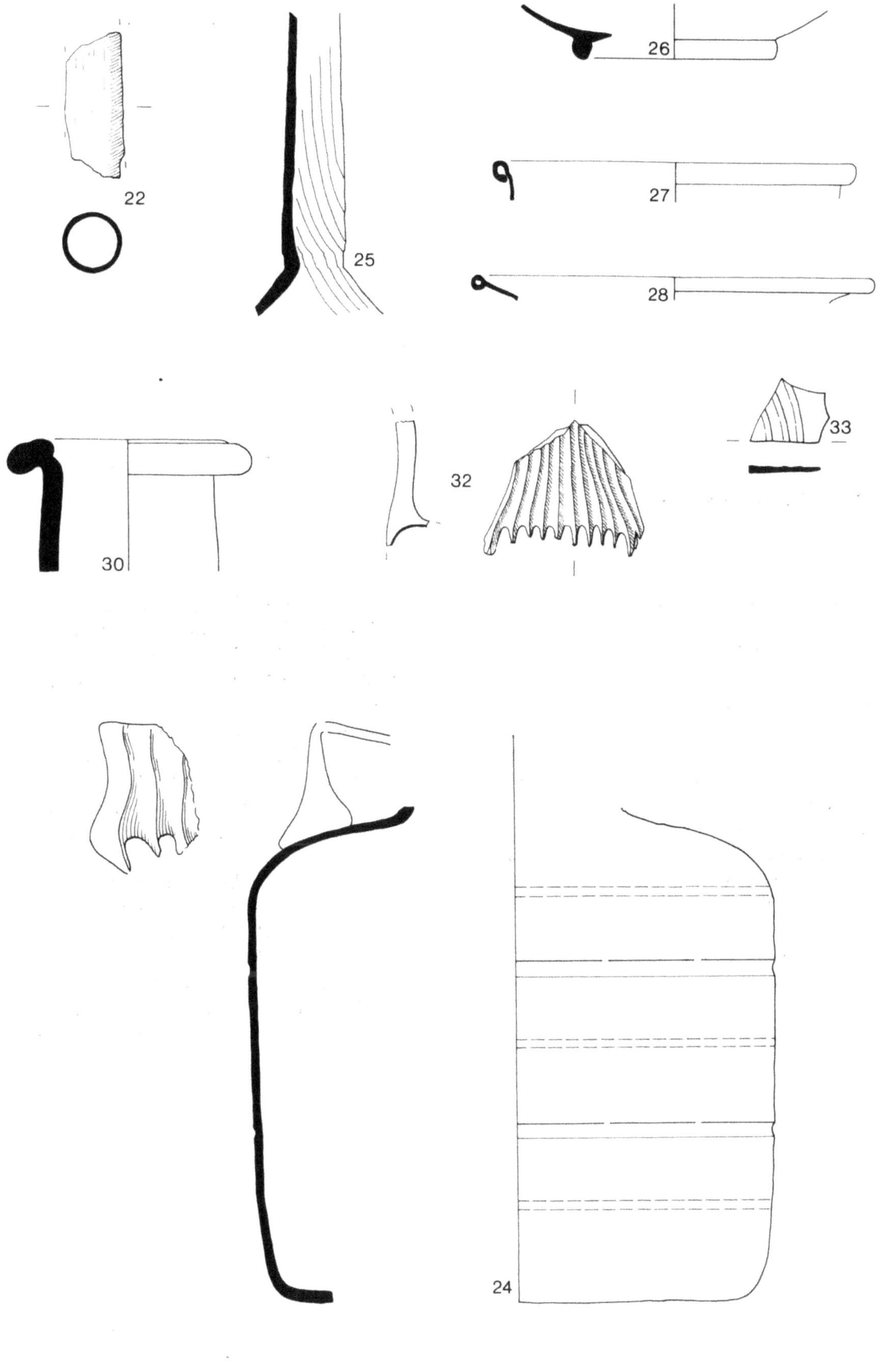

Figure 30: Objects of glass, scale 1:2

(Not illustrated). Eighty-eight fragments were recorded from the following locations: fifteen pieces from topsoil; fifty-seven pieces from post-complex; six pieces from post-Roman; one piece from Stage IV, Room 51A; one piece from Stage VII, Room 29; one piece from Stage VI, Room 51; one piece from Stage VII, Room 51; one piece from Stage VII, north of complex; one piece from Stage VII, Room 33; one piece from Stage VIII, Yard 16; two pieces from Stage VIII, one piece from Stage VII-VIII, west of complex.

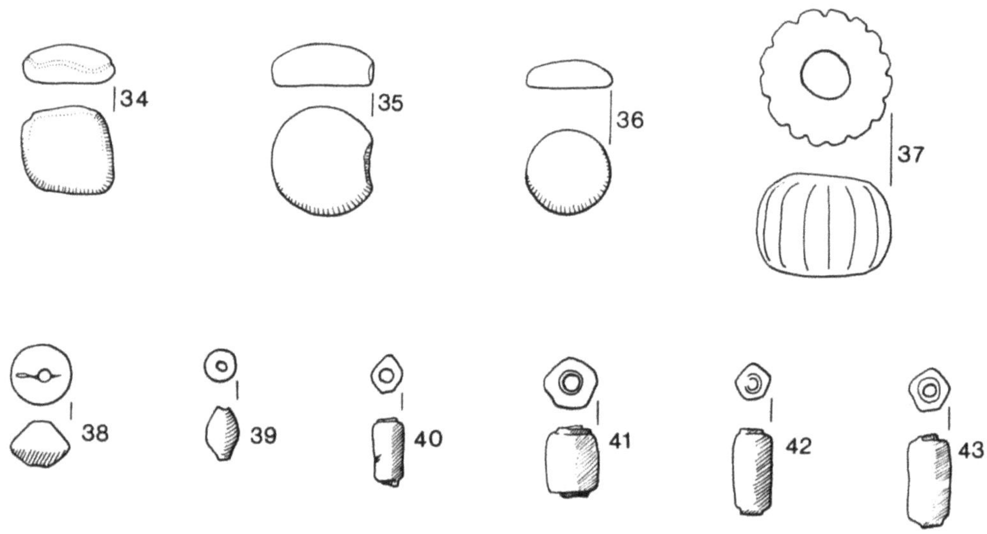

Figure 31: Objects of glass, full size

OBJECTS OF STONE (FIG 32)

By Janet Webster (1976-7)

1. Part of a shale bracelet of approximately D-shaped section. (SF 149, 0/24, post-complex). The curved outer edge is plain but there are traces of ridging on the inner surface. (cf Lawson 1976, 250, 252, fig 4, nos 19-31).
2. Fragment of a stone bowl, probably marble. (SF 379, topsoil above the complex). The elaborately moulded lip could not have been used for pouring, as the rim of the bowl extends in an unbroken line behind the lip. The interior is smooth but the vessel is not likely to have served as a mortar in view of the non-functional spout. Perhaps a garden ornament.
3. Fragments of the rim of a stone vessel, probably of marble. (SF 341, 6/56, Room 39, Stage V).
4. (With P V Webster). Bench support. (SF 354 (upper fragment) 7/56, Room 32, Stage IV; SF 282 (lower fragment) 9/17, Room 40, Stage VII). Two fragments of red sandstone join to form an almost complete support. The front carries decorative carving in the form of an animal leg terminating in a four-clawed foot. The leg is gracefully bent so that in profile the support describes a shallow S-shape. Below the line of the foot the sides of the base are roughly worked, perhaps to allow the base to be set into a floor and cemented in position, although there is no evidence to indicate whether or not this happened. Above the line of the top of the leg the support widens sharply into a rectangular block which extends the full width of the support. The left-hand side is noticeably better finished than the right.
   There seems little doubt that the support belongs to a bench. The better finish of one side suggests that it supported the left-hand end of such a bench which had been placed alongside the west wall of Room 32, constructed in Stage IV (p 29).
   Benches supported by S-shaped marble legs are also to be found at Herculaneum in the *frigidarium* of the Suburban Baths (de Franciscis 1974, 32, pl 28) and the *caldarium* of the women's section of the Forum Baths (Maiuri 1958, 106, fig 83). Similar benches are, however, to be found in other materials and in other types of establishment, as is shown by a wooden bench with S-shaped front legs from a shop on the north side of the Decumanus Maximus in Herculaneum (personal observation).
   The curved profile, sometimes with the lion-claw foot, seems to have been generally preferred for this class of furniture (Richter 1966, 104, figs 528-9).
   Stone bench fragments are not common as site finds but an example from this country may be noted from the *apodyterium* of the baths at Mumrills (Macdonald and Curle 1929, 414, fig 11b and 453; also see Liversidge 1955, 15 for a general discussion of benches in the western provinces).

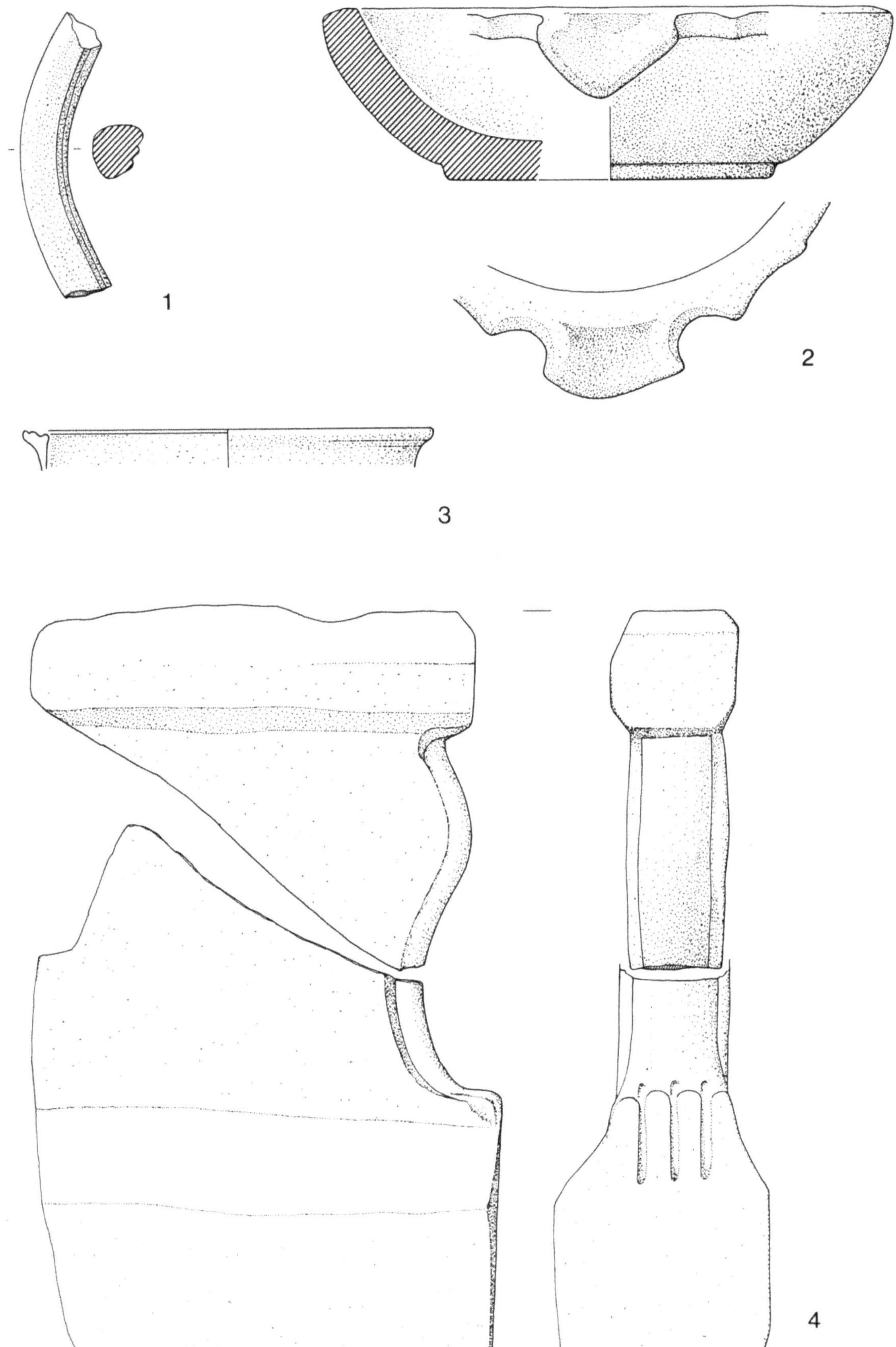

Figure 32: Objects of stone, scale - no1, 1:1; nos 2 & 3, 1:2: no 4, 1:4

THE WALL PLASTER (FIG 33)

By R Ling (1976-7)

Most of the pieces are undecorated. They consist of a brownish-white material containing an aggregate of grit and small brown or black pebbles; grass was used as a binding medium, as is normal in Romano-British plaster. It is not always easy to determine whether the plaster was applied all at once or in successive layers, since such layers often bond tightly together; one may guess from occasional traces and from analogy elsewhere that there were usually two coats. The surface was treated with a whitewash (thickness varying from a mere skim to about 2 mm) and was sometimes carefully smoothed or sometimes left rough with the whitewasher's brush marks on the 'grain' of the plaster (the vertical corrugations formed during the application and drying) clearly visible. On the back there are often the imprints of keying patterns cut in the underlying clay (eg Barbet 1973, 73; 1974, 29).

As well as the undercoat pieces there are a number of pieces with colour washes or patterning. Two groups are discussed below (Decorations A and B) because they are of better quality and show more varied decoration than the rest.

Unlike the plaster from many Roman sites, that from Pentre Farm shows little evidence of redecoration. Only four clear instances have been observed: 3/13 (three successive decorations), 3/22, 7/64, 11/3. All are from rooms which stood for several stages: 3/13 and 3/22 probably derive from Room 19, which was built in Stage I; 11/3 probably derives from Room 11, which was rebuilt in Stage III and stood until Stage VII; and 7/64 probably derives from Room 23, which stood during Stages II and III. Sometimes the redecoration consists of a new layer of plaster over 10 mm thick, sometimes it is a wash barely 1 mm thick. In no case is the preceding decoration 'sparrow pecked' to assist the adherence of the new layer.

The problems of studying the Pentre Farm plaster are common to all sites where deposits are fragmentary and to a large extent in secondary contexts. Many different deposits contain fragments which undoubtedly originate from the same decorations, while conversely single deposits may contain fragments from more than one decoration; but for nondescript white plaster (and sometimes for coloured pieces) the relationships are almost impossible to establish. Moreover, fragments of plaster from a single decoration may vary in thickness and in general appearance according to their position on the wall. Barbet (1973) gives guidance on how to deal with fragmentary finds of wall plaster and discusses variations in thickness. The present analysis, therefore, avoids connecting pieces from different layers unless the affinity is fairly strong. At the same time, the residual fragments may be related to their possible original locations.

The fragments have all been measured to give an idea of the relative quantities involved. The measurements are based on the maximum height and width of each piece - data which will give a reasonably accurate idea of the area represented, since the edges of most fragments are assumed to have been eroded. All figures are quoted in millimetres or square centimetres. A detailed consideration of the two decorations is given below - the catalogue of other fragments being retained in the archive.

Miscellaneous fragments

The plaster fragments have been catalogued and a statistical analysis of Decorations A and B prepared. This data is housed in the archive.

Decoration A (Fig 33)

A more detailed examination is possible for the plaster from 6/51 (found in Room 29 Stage V, but perhaps derived from Room 23), which all seems to belong to one-coloured decoration and is the largest single identifiable complex of wall plaster from the Pentre Farm excavations. Nonetheless, it represents a very tiny proportion of the whole decoration, and there is no question of reconstructing the whole scheme. All that can be done is to distinguish various decorative elements. Only a meticulous search for joins between fragments carrying different colours might shed some light on how the elements went together. In the circumstances, since the range of design and the paucity of joins suggest a representative sample rather than one small part of the decoration, it has been thought worthwhile to make a statistical analysis, which will give a general idea of the decorative emphasis.

The plaster is rather coarse and grey-white in colour, containing a good deal of grit and small dark pebbles; grass was used in the normal way as a binding medium. The surface is generally well smoothed, as is appropriate for a patterned decoration of reasonable quality, and some fragments show clear traces of working with a trowel or other smoothing tool. The smoothing is not, however, always successful: the surface is sometimes uneven and often retains the marks of the brush used to apply the initial whitewash. The thickness of the fragments varies from 9 mm to 33 mm, but the majority measure from 18 mm to 23 mm; many of the thinner fragments have probably split along the joint between two layers, and so do not represent the full thickness of the plaster. On the back face of almost all the thicker fragments are ridges which must be the negative impressions of keying patterns cut in an underlying surface of clay or mortar.

The painting of framing elements is generally careless: stripes, for instance, vary in thickness and intensity and are not always as straight as the designer intended. The leaf and floral motifs are naturally a little more careful; the rosettes and encircling wreaths (Fig 33, A8) have been painted with the aid of compasses and horizontal guide lines.

Fig 33 (A2, 3, 5, 7, 8 and 9) shows the more interesting or informative fragments. The vertical has been judged from the grain of the plaster, but there is no way of ascertaining which is the top and which is the bottom of each piece.

Decoration A: Statistical Analysis (Fig 33)
The following are all from 6/51, Room 29, but perhaps derived from Room 23, Stage V.

1. Plain white fragments. 394.81 sq cm (17.14%)
2. Green, yellow and red motifs (garlands, leaves, fillets?) on white ground. 511.86 sq cm (22.22%), (Fig 33, A2).
3. Borders between pale green and white fields; black or purple lines painted over junction. 174.37 sq cm (7.57%). (Fig 33, A3).
4. Purple stripes and/or red markings on white ground. 100.80 sq cm (4.38%).
5. Black stripes and/or markings on white ground. 135.88 sq cm (5.90%), (Fig 33, A5).
6. Yellow fields with black elements (? leaves or hanging fillets). 204.03 sq cm (8.86%).
7. Borders between yellow and white fields; black lines painted over junction. 142.30 sq cm (6.18%), (Fig 33, A7).
8. Dark red fields with ? white and purple-black elements (rosettes set in wreaths). 429.51 sq cm (18.64%), (Fig 33, A8).
9. Borders between dark red and white fields; black lines painted over junction. 209.95 sq cm (9.11%), (Fig 33, A9).

The analysis, if it has any value, implies that the greater part of the decoration was white ground. One might guess, by study of the fragments and by analogies elsewhere for example Room 2 of the Painted House at Dover (Toynbee 1972a, 260-64; Toynbee 1972b, pl XXVIIIB) that the main (middle) zone of the wall had white fields with coloured surrounds, and perhaps that the upper wall had a continuous white ground. However, several other systems are possible.

Decoration A: Parallels
Decoration A is a modest panel decoration with a few foliate and floral motifs, unexceptional in type, and offering no precise dating criteria. It is not difficult to find approximate parallels on other British sites for the individual motifs. If the black elements in no 6, (not illustrated) are inverted and interpreted as comma-like leaves growing free of one another, rather than as hanging fillets, they resemble the leaves on painted fragments from a villa at Harpham, Yorks (Liversidge 1969, fig 4, nos 6a-c; also, in Gaul, Barbet 1974, figs 22, 58-6, pls IV-VIII). Hanging fillets are common in Pompeian painting, though generally attached to shields or the ends of garlands. The white rosette on a red ground in a circular frame (Fig 33, A8) also resembles a motif from the Harpham villa (Liversidge 1969, 139, fig 4, no 6a), while slightly similar motifs in different colours, but preserving their compass points and guide lines, occur in the villa at Hadstock, Essex (Liversidge 1969, fig 4, no 4a and 4b). Garlands (Fig 33, A2) also appear at other sites, for instance at Witcombe, where a green one and a yellow one seem to intersect (Liversidge 1969, 141-3 and especially fig 4, no 7e).

The most relevant parallels, however, seem to be those at Chester, whence decorators might have been brought to paint at Pentre Farm. Fragments from barrack blocks in the north-east corner of the fortress include several with red, green and yellow floral or vegetal decoration on a white ground, notably a rosette (red) set in a pair of concentric circles (yellow and purple) and part of a green and yellow leaf design which is perhaps the junction of two hanging garlands (the chamfered upper edge of the fragment suggests that it came from beneath a projecting ledge or cornice). Some of the fragments also have guide lines like those used in the Pentre Farm decoration (Newstead and Droop 1936, 38ff, pls IX, X) (I am grateful to Dr G Lloyd-Morgan of the Grosvenor Museum for showing me these and other similar pieces from the excavations in Chester).

Another site in North Wales with foliate wall paintings is Caerhun (Canovium), where a pre-Trajanic room in the *principia* of the fort yielded fragments with a similar colour scheme to that at Pentre Farm, including one with 'traces of a pattern of green leaves with yellow stalks' (Baillie Reynolds 1938).

Decoration B (Fig 33)
This is a distinctive decoration in a close-grained grey-brown plaster with a well smoothed surface; only a few fragments survive, none of them with with a white ground. The fragments are relatively carefully painted and show a wide variety of colours, but there are no leaf or flower motifs: most pieces seem to come from the frames between panels. The sources are: 7/64 (area 380.33 sq cm); 8/31 (26.97 sq cm); 0/3 (50.81 sq cm); 0/28 (5.00 sq cm). Of these the first provides the earliest context (Stage V: Room 36); but, whatever the ultimate source, the plaster is not an original decoration, since it overlies an earlier, much coarser plaster. The thickness varies from 7 mm to 15 mm, but most pieces are between 10 mm and 14 mm (average 12 mm).

Decoration B: Statistical Analysis (Fig 33)

1. Dark brick red with tiny white and yellow splashes. 9.4 sq cm. (7/64 and 0/28, Room 36, Stage V, and Room 31, Stage IV).
2. Series of vertical bands, probably framing elements between panels: brown-purple (at least 25 mm wide); white (8 mm: over brown-purple): brown-purple (11 mm); white (9 mm); dark green (32 mm); white (13-14 mm); brownish red (14 mm); white 5 mm; over black); black with milky white mottling (at least 27 mm). Area 214.60 sq cm.

(7/64 and 8/31 (joining fragments). Room 36, Stage V, and demolition between Stages VI and VII). (Fig 33, B2).

3. Series of vertical bands, perhaps continuous from the last: brownish red or purple (mere trace); white (4 mm wide: over black); black with milky white mottling (78 mm); white (6 mm: over black); red with yellow mottling (at least 30 mm). Total 84.00 sq cm. Further fragments of black with milky white mottling (184 mm) and red with yellow mottling (5.94 sq cm). (7/64 and 8/31 (joining fragments) Room 36, Stage V, and demolition between Stages VI and VII). (Fig 33, B3).
4. A further series: white with pinkish red mottling (at last 22 mm); grey (14 mm); white (5 mm: on grey); grey with white mottling (at least 35 mm). 67.5 sq cm. From 7/64. Further fragment with white stripe and grey with white mottling (5.52 sq cm) from 8/31. (Fig 33, B4).
5. Dark pinkish red. Area 13.82 sq cm. (7/64 and 0/3, Room 36, Stage V and demolition between Stages VII and VIII).
6. Dark pinkish red bordered by vertical black band at least 4 mm wide. 18.4 sq cm. (0/3, demolition between Stages VII and VIII).
7. Dark red bordered by vertical dark green band (at least 14 mm wide) carrying traces of white stripe. 7.14 sq cm. (0/3 demolition between Stages VII and VIII). (Fig 33, B7).
8. Reddish brown and creamy yellow zones with vertical black stripe 6/7 mm wide) painted over junction. 12.95 sq cm. (0/3 demolition between Stages VII and VIII).
9. Indeterminate colour. 4.83 sq cm. (7/64, Room 36, Stage V).

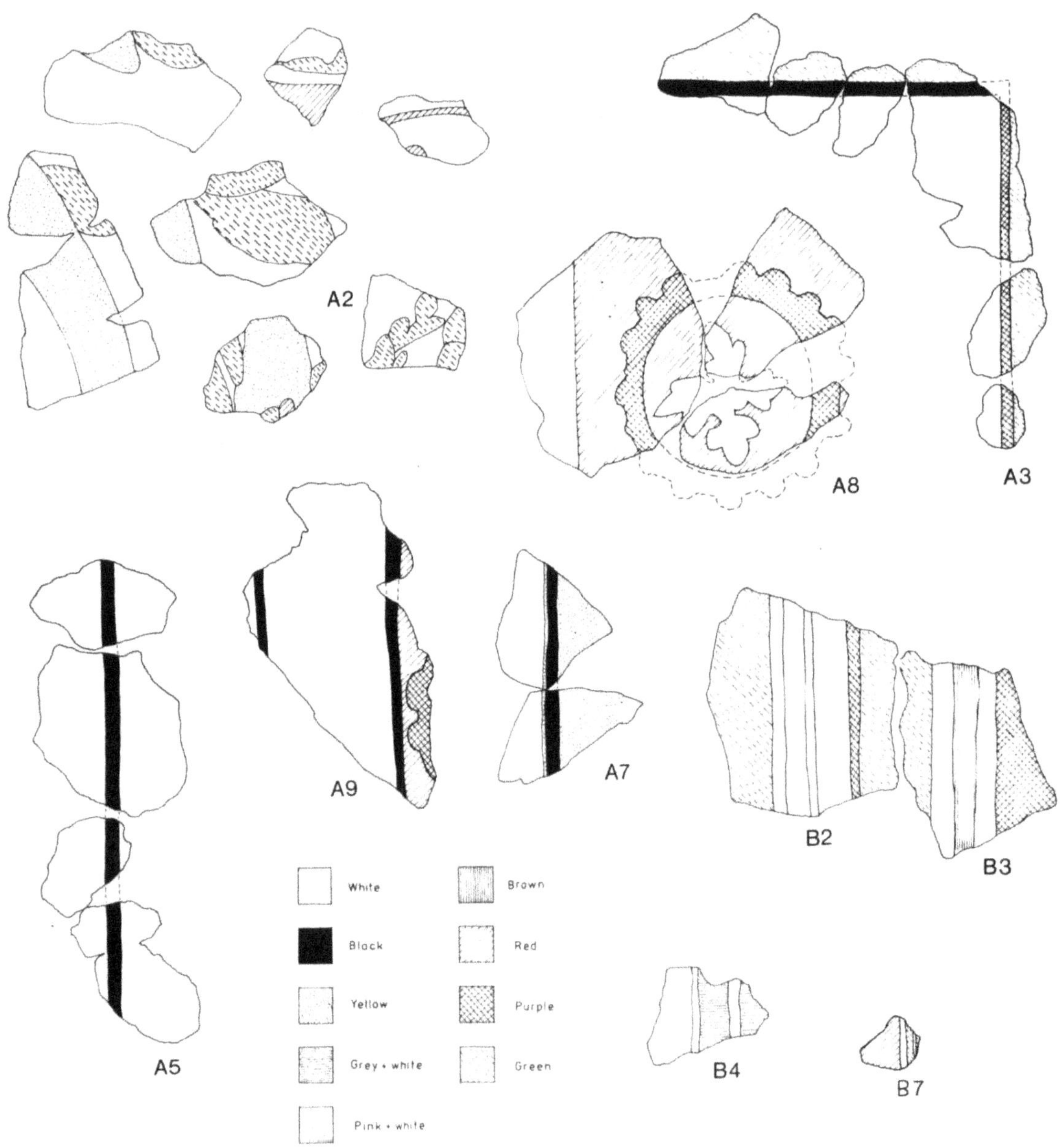

Figure 33: Painted wall plaster, scale 1:3

BRICK AND TILE

By P V Webster (1976-7)

Pentre Farm yielded large quantities of brick and tile. Overall quantities from each layer have been noted and the layer-by-layer details deposited with the site archive. A summary by stage and room is given in Table 3.

Categories of brick and tile identified

Roof Tiles

Both *tegulae* and *imbrices* were present in sufficient quantities from later stages to make it virtually certain that they formed the method of roofing. A few *tegulae* showed signs of modifications; such modification, if it takes place after firing, is difficult to detect unless substantial pieces of the tiles are found or they are *in situ*. However, at least one *tegula* had been modified during manufacture to reduce the flange to a mere tenon 50 mm long. Insufficient remained of the tile to ascertain whether more than one tenon was present. Modification of roof tiles for use elsewhere in the construction of a building is well attested (cf Blake 1947, 305 and *passim*). However, the most likely use of the present tile would be in the uppermost row of tegulae on the roof, the flange having been cut away to accommodate either the ridge or a decorative finial. One *tegula* with a hole for a nail or peg was also noted, perhaps part of the lowest course of such tiles which would have been nailed or pegged in place (cf Blake 1947, 304; Brodribb 1987, 10-11).

Box Tiles

As is normally the case, these probably formed the means of providing wall flues for the bath-house hypocausts. All true box tiles appeared to be scored for cement or plaster on two opposing sides, with vents for communication between the various flues in the remaining 'plain' sides. Scoring had been done either with a point in a lattice design or with a comb (usually in a wavy-line pattern). No complete boxes were found, nor could any be reconstructed from the examples submitted for study, but widths of 13-16 cm for the scored face were recorded and the plain face appears to have been of similar dimensions. The placing of vents was not entirely standard; vents were *c* 9-10 cm long and *c* 3.5 cm wide where measured and were placed so that they began *c* 6-10 cm from the end of the tile. If the vents were symmetrically placed as seems probable then the overall length of the box tiles would have been *c* 25-30 cm.

In addition to the true box tiles a number of aberrant forms were noted. Some box tiles appear to have been made and then cut in half before firing so as to make L-shaped or gable-shaped tiles. The side vents of the box tiles were retained in this form. These tiles may have been used to protect wooden beams within the bath-house structure. No signs of a concrete vaulted roof were found, so it may be assumed that the roof was of timber. Some means of protecting the roof joists from heat and moisture would have been required and a suspended tile ceiling is possible. Encasement of the main beams with ceramic boxes would have formed an additional protection against the heat and damp of the hottest rooms.

A few tiles contained circular holes *c* 50 mm in diameter on the plain sides. Too little remained to ascertain if they were true box or L-shaped tiles. It is possible that these are merely box tiles with a slightly different form of vent. However, a similar function to the L-shaped tiles is also possible (see Brodribb 1987, 76, fig 33, c).

Floor Tiles

Large tiles *c* 2.5-3.0 cm thick. These were generally scored in an open lattice pattern on the underside, no doubt to aid adhesion to a mortar matrix. They were found *in situ* in the lower or sub-floors of the bath-house hypocausts, but could have served other purposes as well. They correspond to the 18-inch tiles *(Sesquipedales)* recommended by Vitruvius (*De Architectura*, Vol 10, 2) for the purpose. They would also have formed the ideal capping to the hypocaust *pilae*, acting as a base for the mortar-concrete floor of the hot rooms. Another function would have been to act as a ceiling to the hot rooms where they could be fixed with the aid of metal clamps and plastered to protect the roof from heat and damp (cf Blake 1947, no 310 and *De Architectura*, Vol 10, 3 also Brodribb 1987, 3, for brick and tile names and sizes).

Bricks

Two sizes of brick were recorded: 27.3 cm x 27.7 cm x 4.7 cm (weight 6 kg) and 19.5 cm x 19.5 cm x 7.5/8.0 cm (weight 5.45 kg). The smaller correspond to the *Laterculi Besales* of classical literature and practice (Blake 1947, 306; Brodribb 1987, 34-6) and were the recommended components of hypocaust *pilae*. The others appear to be larger versions of the same type (cf Blake 1973, 109 for the use of this size in the Antonine period) or perhaps *pedalles* (Brodribb 1987, 36-7). The larger bricks probably formed a slightly wider base and cap to the *pilae* although other uses for both types within the structure are possible.

In addition to the material described above there were large quantities of brick and tile from unstratified levels, mainly, no doubt, the result of the final demolition of the bath-house. Levels not associated with the building yielded very low quantities of brick and tile and have not been included in Table 3.

The notable absence of brick and tile from levels prior to Stage IV presumably indicates that the first structures on the site were roofed with some material other than tile.

Vents

One fragment of chimney pot was found and is illustrated (Fig. 45, no 329). It is not possible to offer a reconstruction of the complete chimney from such a small piece. The possible range of chimney forms decorated with thumbed zones is considerable (Lowther 1976, figs 1-4).

Table 3: Weight of tile from the complex

Stage	Room (weight in kg)	
Demolition between III & IV	29 (0.23), 31 (3.35)	(3.58)
IV	30 (0.43). 33 (2.68)	(3.11)
IV alterations	33 (4.05)	(4.05)
Demolition between VI & V	19 (0.45), 31 (0.80)	(1.25)
V	36 (0.18), 39 (22.35), 51 (0.20), 7/35 (18.55)	(41.28)
VI	7 (14.68), 30 (1.33), 46 (3.85), 50 (3.18), 51 (286.63), 54 (1.20), 7/38 (22.23)	(333.10)
Demolition between VI & VII	11 (12.4), 12 (2.85), 33 (6.08)	(21.33)
VII	11+12 (6.76), 30 (4.38), 32 (2.23), 36 (24.66), 38 (3.04), 39 (6.83), 40 (1.78), 42 (92.01), 43 (4.80), 44 (56.33), 45 (44.47), 46 (15.14), 49 (0.30), 51 (2.90), 52 (25.88)	(291.51)
VIII	42 (3.13), 46 (3.30), 49 (1.26), 52+53 (9.6)	(17.29)
VIII alterations	29 (0.66), 46 (12.31), 49 (0.65)	(13.62)

STAMPED TILES AND GRAFFITO

By M W C Hassall and R S O Tomlin (1976-7)

Stamped tiles

(Not illustrated). Fragments of nine stamped tiles were found in 1977 during excavations of the Roman bath-house. All are probably from different dies and appear to have read LEG XX VV, *Leg(io) XX V(aleria) V(ictrix)*, except one, similar in style to Holt no 3 (Grimes 1930, fig 59) which may have read LEG XX VAV, for *Leg(io) XX V(aleria) V(ictrix)*. Of the rest only three can probably be matched at Holt (Holt nos 2, 6 and 7 respectively); two cannot be matched although one is set in an ansate panel similar to that of Holt nos 5 and 8; the remainder are fragmentary but are also probably from dies different from the published Holt examples.

Graffito

(Not illustrated) Fragment of box flue-tile 105 mm by 75 mm, 5 mm thick (incomplete). Below the bottom edge of the square opening originally cut in the side of the tile is a graffito which has been cut by a practiced hand before firing: **QVINTVSΛVII[...] ΛMΛBIILISV[...] [VS ΛNTI[... ...]V[.** In line 2 the last letter could be X. The graffito could represent the *tria nomina* of two individuals, as Quintus Avid(ius?) Amabilis, A[u] lus Anti[stius? Feli]x. The rare *praenomen* Vibius, though theoretically possible as an alternative to Aulus, is much less likely.

MAMMALIAN PAW PRINTS ON CERAMIC TILES

By M A Edington (1976-7)

1. (Not illustrated). The small foot resembles that of a modern domesticated cat (*Felis domesticus*) rather than the indigenous wild cat (*Felis sylvestris*). The presence and status of domesticated cats in Britain in Roman times are matters for debate and poorly documented. (8/7, demolition debris, Rooms 11 and 12, Stage VI-VII).
2. (Not illustrated). From the size of these dog tracks, the animal would have been approximately the size of a modern fox terrier. Dogs of this size would have been used for hunting; there is also evidence from mosaics, etc that dogs of this size were kept as pets. (8/12, Room 45, Stage VII).
3. (Not illustrated). The tracks indicate that the animal, a medium to large dog, walked well forward on its toes, and thus that it was bred for speed, like the modern greyhound. It would seem likely that it was of similar size to a small greyhound (though probably rough-coated) and was used for hunting. The closest analogy would be the lurcher, bred in more recent times for hunting rabbits. (Topsoil above complex and 0/13).
4. (Not illustrated). A slightly smaller dog than suggested by no 3 above (may be a female of the same type). The animal was trotting when the tracks were made, with a consequent overlap of the marks. (Ploughsoil).
5. (Not illustrated). The tracks could not be identified as those of a dog or cat. The nearest track is that of a large mouse. (0.25, medieval robbing).

Part III: The pottery

by P V Webster

with
D Bailey, B M Dickinson, B R Hartley & K F Hartley

INTRODUCTION

The report which follows was originally written in 1976-7. Further work on the site records subsequently has necessitated a re-ordering of the text and this has been followed by its revision in 1987. Time has not, however, allowed for major rewriting and it remains essentially the report as originally submitted.

THE SAMIAN POTTERY

Little samian ware was recovered from the site; what was found consisted mainly of small sherds and was often residual or from unstratified contexts. The majority of the more substantial fragments of decorated samian has been included in the catalogue below and a selection illustrated to demonstrate the range. Of the plain samian only one flagon (no 17) has been discussed in detail, because of its intrinsic interest, but apart from this the plain ware formed too small a sample in each stage to add significantly to the chronology of the site. Some of the samian evidence has been summarized in an unpublished chart which formed the basis for part of the numerical analysis presented in the section on trade links and social implications.

Any conclusions are necessarily tentative. However, the general date range of the material is clear. South Gaulish pottery is present in significant quantity, but the forms are predominantly late. There is only one example of form 29 and that a very coarsely rouletted rim which could well date from the very end of the period of production of the form. There are only a few examples of form 18, and the sherds of form 37, where decorated, are all late in the South Gaulish series. On this basis it seems possible to suggest that occupation is unlikely on or near the site before *c* AD 90.

For the latest date of occupation we shall have to rely on the coarse pottery evidence. It may be noted, however, that later Antonine forms such as 79/80, 45 and Curle 21 are present in small numbers.

THE COARSE POTTERY

Coarse pottery formed the largest single finds category at Pentre Farm and, as well as dating the site, it provides much of the basis for social and economic considerations.

Analysis of the pottery began on site where all sherds were counted and sorted into rims, bases and body sherds as well as into broad fabric categories. However, these have not been used as the basis for the kind of quantification presented by Fulford (1975) for three reasons: firstly, the fabric divisions are not sufficiently detailed; secondly, the high number of structural stages divides the assemblage into groups which are too small to be statistically viable; and lastly, there is a high proportion of residual pottery in each stage. In sum, both the size and composition of the sample suggested that statistical analysis was unsuitable. A general assessment of pottery distribution was made as part of an attempt to establish functional differences between rooms.

It was decided that the best overall picture could be obtained by illustrating and/or describing all stratified rims and this has been done for all stages except the latest (layers above the demolition work of Stages VII and VIII). Body sherds have also been mentioned where they seem particularly relevant (for example, Nene Valley colour-coated ware is represented almost entirely by body fragments). The aim has been to present as complete a picture as possible of the ceramics from each stage and to make possible a minimum vessel count at some future date, if so desired. This has been done mainly through the catalogue (pp 99-120).

Sources and Fabrics

Although it is in Wales, Pentre Farm is geographically part of the Cheshire Plain and it is very similar to other Cheshire Plain sites in its ceramic assemblage. Two major groups of sources for coarse pottery other than mortaria can be identified for Pentre Farm. These are the several centres producing Cheshire Plain wares and the Black-burnished ware industry of Dorset.

Cheshire Plain Fabrics. The products of all the known kiln sites on the Cheshire Plain (ie Holt, Wilderspool and Northwich) share certain characteristics. The principal fabrics present on the sites contain a high proportion of sandy filler, no doubt derived from the sand deposits which are a feature of the surface geology of the region. The exact proportion of sandy filler used in any particular pot seems to vary but this will not necessarily help distinguish the product of one kiln from that of another. Variations may occur because of factors such as the stiffness of the clay and weather conditions: that is, the amount of filler needed for any given pot will be different on a cold wet morning from that required on a hot dry afternoon. Furthermore, the colour of the fabrics found at each centre is generally fairly similar. All centres were probably producing both oxidized and reduced fabrics mainly within the orange through orange-buff to light red and through light to mid-grey. In this situation it is often difficult to distinguish the products of different kilns, so the term 'Cheshire Plain fabric' has been used here to characterise these sandy fabrics. The parallels quoted for examples of the fabric will often be from a known kiln site, but this does not necessarily imply that the vessel in question was made at the site and indeed it is unlikely that all the Roman kilns in this area have been discovered.

Of course, certain products from the Cheshire Plain can be distinguished. The fine wares made at Holt, for example, are very distinctive (Greene 1977) and several examples of these found their way to Pentre Farm (nos 59, 118-19, 306). Even within the more uniform Cheshire Plain fabrics certain forms seems to be characteristic of known kilns. Thus, flanged bowls produced at Holt appear to be characterized by two grooves, one towards either edge of the flange (Grimes 1930, nos 83-6), while Wilderspool appears to have specialised in rough-cast and rouletted beakers and a distinctive form of bowl, sometimes thought to be reminiscent of the samian form 44 (Hartley and Webster 1973, 83, nos 35-7; 85, nos 53-5). Where forms are thought to originate at certain kilns this is stated in the catalogue, although it must be borne in mind that other kilns producing similar forms may yet await discovery in the area. For further discussion of Cheshire Plain Wares see Webster 1982.

Black-burnished Ware. All Black-burnished Ware from the site falls within that branch of the fabric known as Category 1. The source of at least some Category 1 vessels seems likely to be Dorset (Williams 1977) but other sources are possible (among a small number of Category 1 samples from Wilderspool examined at least one non-Dorset fabric appears to be present: Williams and Hinchcliffe, forthcoming). It would seem that the first appearance of Black-burnished Ware in quantity on the Cheshire Plain, as further north, was a Hadrianic phenomenon. The evidence from Northwich (Jones *et al* 1971) is fairly conclusive; the samian evidence shows that the two periods on the site can be divided into a late first to early second century, Period I (without Black-burnished ware) and a Hadrianic to early Antonine, Period II (with Black-burnished Ware). There thus seems no good reason to suppose that the date of *c* AD 120 proposed for the introduction of BB1 in quantity to the Wall zone should not also apply to our area.

The types of Black-burnished Ware present at Pentre Farm closely resemble those found throughout northern Britain (Gillam 1976 gives a fairly complete range of types found in this area). It may, however, be noted that several examples of lids in Black-burnished Ware occur. The Black-burnished lid seems largely to have escaped mention in the literature on the ceramics of northern Britain, despite the appearance of a complete example at Holt (Grimes 1930, 155, no 100). A preliminary review of the available evidence will be found in Wallace and Webster forthcoming. Such a simple form could well have a wide date range but most of the examples known to the author are from second century contexts and it is worth noting that such a lid would be suitable for use with the flanged bowl of the second century and would convert the latter into a casserole at least as well as the flanged dishes suggested for this purpose by Gillam (1976, fig 6; nos 90-91).

Miscellaneous fabrics

Colour coated white wares. There are a few examples of this fabric in later levels but, as one would expect on a predominantly second century site, the ware is not common. Some ware will have been imported from Cologne (see Anderson 1981) some from the Nene Valley.

Severn Valley Ware. Again, there are only a few examples and Pentre Farm is clearly on the extreme edge of the known West Midlands distribution (Webster 1976, figs 11-12). It may be significant that the ware first appears at Pentre in the latest phase (Stage VIII). This would fit in with the pattern seen at Manchester (Jones and Grealey 1974, 94), in contrast with the earlier appearance of the ware at Chester (Carrington 1977b).

Sources of mortaria (K F H)

Twenty-two identifiable mortaria were examined. The main suppliers in the Flavian period were probably the makers of form Gillam 238, worked in Gallia Belgica, perhaps in the Pas de Calais. Their mortaria often occur in greater numbers on sites near the coast, perhaps as a result of coastal distribution. There seems also to have been a local source already functioning before the end of the Flavian period (nos 71, 314, possibly from one vessel). About half of the sample, can be dated within AD 100-160+ and was supplied from local sources in Cheshire (including Wilderspool) and North Wales (eight and three examples respectively). The single example of form Bushe-Fox 26-30 (AD 80-150) could well belong to this period. The remaining seven mortaria cover the period AD 150-300/350, indicating a remarkable reduction in the use of mortaria on the site, only partly explained by the general chronology. All but two of these were made in the Mancetter-Hartshill potteries which obviously took over this tiny market when local supplies failed. The Oxford potteries only reached their peak of production in the second half of the third century and, probably for this reason, made little impression here (one example only). The single example from the Rhineland is not surprising, since such mortaria are found in small numbers dispersed throughout Britain, on sites occupied within the period AD 150-250.

It will have been seen from the discussion of sources above that the major suppliers of mortaria were the centres at Mancetter-Hartshill and near Wilderspool. For convenience and to avoid repetition these two major mortarium fabrics are described below.

Mancetter-Hartshill. A hard, fine-textured white or cream fabric sometimes with slightly darker slip. It contains very little tempering and the trituration grit used after c AD 140 is composed of red-brown or dark brown material (usually pre-fired waste pottery), often mixed together.

Stockton Heath, near Wilderspool. The fabric commonly produced here was a soft, basically fine-textured one, orange-brown in colour, sometimes with pale bluish-grey core; the fabric can be so heavily tempered with sandy particles that it is often abrasive to the touch. The mortaria may be self-coloured or have a cream, white or red-brown slip, which does not survive well. The trituration grit is usually a mixture of quartz and red-brown and brown material. None of the mortaria amongst the waste from the kilns has all quartz trituration grit, which is normal at Holt and in many mortaria likely to have been made in Wales. It is, however, not yet possible to be completely certain that quartz grit was not used alone at Wilderspool, though it seems unlikely. (A fuller description of the fabrics of mortaria made in these potteries can be found in Hartley and Webster 1973, 91). This was also one of the potteries producing the so called Raetian or sub-Raetian mortarium. (cf Robertson 1975, 241-44).

The amphorae

The study of the amphorae from Pentre Farm seems more likely to shed light upon the trade contacts of the site than on its chronology. In the hope of obtaining an indication, not only of the amphora-borne commodities reaching the site but also of their relative importance, all amphora sherds were weighed and divided according to commodity, stage and room (details retained in archive). This information is summarised in Table 4, which follows a brief description of the amphorae types present. Illustrated amphorae appear in the catalogue as nos 323-7. For fuller discussion of the identified forms see Peacock & Williams 1986.

Types present (identifications by D P S Peacock)

A. Amphorae of Dressel form 20 containing olive oil from the area of South Spain between Cordoba and Seville. First to second century AD.

B. Amphorae of Pélichet form 47 containing wine from South Gaul. Mid-second to third century AD.

C. Amphorae of Dressel forms 2-4 containing wine from Campania.

D. Possibly one sherd of an amphora of Camulodunum form 185 containing Garum from South Spain.

E. A handle fragment perhaps from a Rhodian-style amphora with peaked handles. The handle is round in cross-section and of fairly coarse light orange-red fabric, burnished on the outer handle. Commodity unidentified, but probably wine from the Aegean.

F. Fabric unidentified. Sherds of this fabric may well be from a globular amphora. The fabric was, however, smoother than that used normally for Dressel 20. The fragments recovered were cream-buff with a grey core and included fine white and grey inclusions and some mica. The surface was smooth and originally had a cream slip.

G. Fabric unidentified. A single sherd, apparently also from a globular form. The fabric is sandy with some mica. The colour is predominantly pink-buff but with a rose-pink inner surface and rose-pink immediately below the outer surface.

H. The fabric (unidentified) is a hard dusky pink with few visible inclusions but some mica present. The surface has a light cream slip. The extant fragments (no 323) are illustrated. They are clearly from a relatively narrow amphorae with a narrow pointed base.

Amphora-borne commodities were reaching Pentre Farm from at least eight sources in the Mediterranean. To assess the relative importance of these sources the amphorae were quantified by weight. The globular Southern Spanish oil amphorae (Dressel 20) are shown to predominate even allowing for the greater weight of this type compared to other amphora types. 28.6 kg of Dressel 20 were present in comparison with a total of 1.45 kg for all other types. The wine amphorae from Southern Gaul (Pélichet 47) are next in importance while other sources are represented by only a few sherds each.

One other point suggested by Table 4 is that amphorae were only present in quantity during the later stages. Most amphorae are from Stage IV or later and the majority from Stage VII or later. This may have implications of function or trade, there is no reason why Dressel 20, in particular, should not have been traded in this area at almost any time during the history of the site.

Fabric	Pre-Complex	III	III-IV	IV	IV-V	V	VI-VII	VII	VII or VIII	Post VII	VIII	Post VIII	Post Complex	North of Complex	Weight (kg)
A	1	2	3	6	1	1	5	16	4	64	9	7	?		28.60
B									1	3	1				0.65
C					1					1					0.15
D										1					0.05
E											1				0.05
F										1					0.25
G												1			0.05
H														1	0.25
Total	1	2	3	6	2	1	5	16	5	70	11	8	?	1	30.05

Table 4: Summary of amphora fabric types from all phases

TRADE LINKS AND SOCIAL IMPLICATIONS

The trade links revealed by the Pentre Farm coarse pottery have, in the main, already been outlined above. The major sources of all coarse pottery were local but there were also important contributions from outside the immediate area: Dorset for some or all of the Black-burnished ware and Mancetter-Hartshill for some of the later mortaria, as well, of course, as South and Central Gaul for the samian ware. Most other sources are quantitatively unimportant but nevertheless of interest. Mortaria were also being drawn from Kent or North-east Gaul and Oxfordshire; other coarse pottery came from the Severn Valley, the Nene Valley, Lezoux and the Rhineland. Lamps were imported from Gaul and, almost certainly, from Italy. The number of these sources, which are minor ones in terms of this particular site, is not large but the geographical range is great. The small quantity of such wares present at Pentre Farm is further evidence for the existence of middlemen in the Romano-British pottery trade, as the marketing of such small numbers of vessels would be uneconomic, except as part of the mixed stock of such a man. It might also suggest the market-pull of the site or the proximity of roads.

Imported amphorae also form part of the overall trade pattern. Globular amphorae holding oil from South Spain predominate, as one would expect, but other vessels present indicate the importation of wine from South Gaul and Campania and of fish products from South Spain. A number of amphorae of unidentified source suggest that the list is not exhaustive and that other commodities from other parts of the Mediterranean were reaching the site.

Some measure of the social status of the site and its occupants may be gauged by the more 'exotic' items present. The three lamps (nos 24, 173, 221) clearly fall into this category. Imported lamps are not common as site finds in northern Britain, and second century Italian lamps (nos 24, 221) are not common in Britain as a whole (pers comm D M Bailey). Similarly, the Holt fine wares (nos 59, 118, 119, 306) are rare even in the immediate vicinity of the kilns and far from common at Chester itself (pers comm P Carrington). Both categories of fine ware, therefore, imply a certain wealth. It may be significant that both the lamps and the Holt wares are unlikely to have reached the site after the Hadrianic period, ie that both belong to the earlier stages in terms of their original appearance on the site. The contrast with the later stages is marked. On a site which could afford imported lamps one would expect British and Continental colour-coated fabrics if its status were maintained into the appropriate production period, but despite occupation into the mid third century Pentre Farm yielded few examples of Nene Valley ware and only single examples of Lezoux and Rhenish colour coats.

The samian ware shows a similar decline in quality when the earlier and later stages are compared. The great majority of the decorated pieces, for example, belong, or appear to belong, to the first half of the second century. However, the overall quantity of samian is low, and even if the earlier stages of the site were comparatively wealthy in comparison to the later, the overall picture does not seem to indicate particularly great wealth or high social status at any period.

Qualitatively then the later levels seem poorer and quantitatively the same would seem to be true. The overall numbers of later second and third century pieces are relatively small. This may well reflect a change in the occupation of the site.

In summary, the pottery suggests an economic and perhaps social decline on the site after an initial period of relative prosperity. It is not possible to ascertain from the ceramic evidence when the apparent initial prosperity ceased, although economic decline from the mid second century onwards is evident from the structural sequence (between stages III and IV).

CHRONOLOGY

At Pentre Farm the eight stages of development have to be accommodated within a comparatively narrow timespan. The number of sherds which are of use in dating the structural sequence is not as high as one would wish and the dating of a stage is often heavily dependent on relative chronology.

Stage I.
The samian evidence suggests occupation near the excavated site from *c* AD 90, and is supported both by the large number of late first to early second century coarse pottery vessels present on the site, and by the date of the material from earlier excavations at Pentre Ffwrndan (see below p 120). The construction of the complex cannot, however, date from as early as AD 90 because amongst the small amount of material from Stage I were sherds of Black-burnished ware (nos 26-7) and Central Gaulish samian (not illustrated) of form 35. Black-burnished ware is unlikely to have been imported into the area in quantity before *c* AD 120. Therefore, if nos 26-7 do belong to the initial construction of the complex, it was probably built after *c* AD 120. An early Hadrianic date for Stage I may, therefore, be suggested.

Stage II.
The latest pieces from this stage (nos 32, 35-6) are Hadrianic-Antonine but none need necessarily be later that Hadrianic. The assemblage is comparatively small but seems similar to that found on military sites thought to have been abandoned at the time of the Antonine advance into Scotland.

Stage III.
There is only a small amount of pottery from this stage, but it may be considered together with that from the Stage II-III midden (nos 37-47) and the sherds from the demolition between Stages III and IV (nos 58-69). The latest pieces are from a constructional contexts (nos 51-3) and date to the mid to late second century. No 64 from demolition is late second-early third century. These suggest that this stage may be of Antonine date.

Stage IV.
This stage includes some Antonine pieces (nos 80, 85, 90) and a number which are either later Hadrianic or Antonine (nos 86, 94). A late second century date is likely.

Stage V.
Colour coated white wares from Cologne or the Nene Valley (no 136) first appear in this Stage. Nos 109 and 112 are also likely to date from the second half of the second or early third century.

Stage VI.
The latest pieces from this stage are nos 138 of late second-early third century (from a construction or immediately post-construction layer) and 147 and 150 (from occupation levels), both of mid to late second century. The type of dish with developed intersecting arc decoration does not seem to be present in Antonine Scotland and a date of introduction prior to *c* AD 170 seems improbable. Stage VI should, therefore, date to the late second or early third centuries, or later.

Stage VII.
The high proportion of residual material in this stage may be noted. No 190 is, however, probably early third century. There are a number of much later pieces in the demolition deposit (nos 230-33, 250) and these probably reflect later activity on or near the site. The majority of the pottery from the site should or could pre-date *c* AD 260.

Stage VIII.
The material from Stage VIII may be grouped with that from the VII-VIII accumulation in the ditches to the west of the complex. No 260 is probably mid third century or later. There is nothing to suggest definite Stage VIII activity after the early to mid third century.

Unstratified.
The unstratified material includes a number of later pieces but differs little from the picture already offered above. Sherds which are definitely later than *c* AD 240 are very few indeed and the end of major occupation on the site may have occured at about this time, although some sort of activity on reduced scale is implied by the few later pieces present and by the pit-fill 6/21 (nos 294-8).

CONCLUSIONS

The ceramic evidence is too imprecise to allow any exact dating of the structural stages at Pentre Farm. The total timespan, however, is not large: settlement near the site is unlikely to have occured outside the period *c* AD 90 - *c* 300 and the building complex itself seems to belong to the period *c* AD 120-240. On the evidence offered, it seems possible to suggest the following dates:

Stages I-III early to mid or late second century
Stages IV-VII late second to early third century
Stage VIII early to mid third century

CATALOGUE

Soil conditions appear to have had some effect on the coarse pottery from Pentre Farm. Surfaces are often somewhat eroded and it is not always possible to describe original surface treatment. Most of the pottery is fairly hard (ie it offers some resistance when scratched with the finger nail, 2.5 on Moh's scale of hardness, and a fabric is only described as soft or hard when it varies markedly from this norm).

A number of the more common fabrics have been described in the introductory sections. Other fabrics (eg Black-burnished ware, Severn Valley ware) are adequately described in the literature referred to in the text. All other fabrics can be considered to contain some filler unless otherwise stated. All fabrics which are coarser than Black-burnished ware Category 1 are described as 'coarse'.

All items are from the complex unless otherwise stated.

The samian stamps (Fig 34) (BRH & BMD, 1976-7)

The following superfixes are used.
1. Stamps recorded from the factory in question.
2. Other stamps of the same man recorded there.
3. Assigned to the production centre on the grounds of fabric and/or distribution.

1. Form 27 stamped **BIRA(C)ILLI.** Biragillus ii 7a. Les Martres-de-Veyre[1]. The stamp was only used on cups, and occurs in the groups from the London Second Fire. Since other stamps of his come from Catterick and Ilkley, he clearly worked *c* AD 100-25. (0/12 and 0/80, post-demolition between Stages VII and VIII).
2. Form 27 stamped **FCO.IV.A?)** Co(sius?) Iucundus 4a. La Graufesenque[3]. OFCO.IV was clearly intended, and it is possible that the stamp is from a broken die. The potter's record is basically Flavian, and includes sites such as Caerleon and Newstead. *c* AD 75-100. (10/75, demolition between Stages III and IV).
3. Form 27 stamped **DO CF**. Donnaucus 5a. Les Martres-de-Veyre[1]. Donnaucus was one of the potters whose work (including this stamp) appears in the London Second Fire groups. This particular one has also been noted from Malton (presumably before AD 125). *c* AD 100-25. (From topsoil above the yard north of the complex).
4. Dish stamped **OFP V I.** Paullus iii 3a. La Graufesenque[2]. This stamp is known from sites such as Caerleon, Catterick (two examples) and the Saalburg. Other stamps appear at Corbridge and Wilderspool, but as there is also one from Risstissen and, as he made form 29, a date *c* AD 70-95 is likely. (From topsoil above Room 28).
5. Form 27 stamped **[P.AVL]VSM** retr. Paullus iv 10a. Lezoux[2]. This is one of the less common stamps of the potter whose decorated ware has associations with the early work of Cinnamus ii. It occurs in the Birdoswald Alley group. His plain ware stamps are otherwise known in the groups from the Castleford Pottery Shop (of the 140s) and Verulamium Second Fire (*c* AD 155-60). *c* AD 135-160. (4/22, from the west of the complex, Stage VII of later).
6. Form 38 stamped **SAC[EROM]** (III-I). Sacero la. Lezoux[3]. This stamp was used on forms 18/31-31, 18/31R, 27, 31, 33 and 38, and there is one example from the Newton Kyme *vicus*. Sacero may only have used this one die, but there are reports (unverified) of another stamp from Birrens and one on form 79. It was used on the rim of form 37 with an ovolo probably by Albucius ii and is on a burnt bowl from the Wroxeter forum. *c* AD 150-80. (From topsoil above the bath-house).

Decorated samian (Fig 34) (PVW, 1976-7)

7. Form 37. South Gaulish. Basal portion and some flakes, showing panel decoration separated by wavy-line borders. The panels contain from left to right: the feet of a (?)dancing figure over stylised grass; the dog O.1994; a pair of feet, probably part of the Diana, O.104. The two latter figures occur together on a vessel from La Graufesenque (Hermet 1934, pl. 87, 3). *c* AD 80-110. (4/1 and 0/42, west of the complex, Stage VII or later).
8. Form 37. South Gaulish. Four joining pieces showing an ovolo with triple tipped tongue above a wavy line. The decoration shows the boar, O.1670 and part of the dog, O.2039W, separated by a stylised bush. Below is a basal wreath composed of triple leaves (Knorr 1919, 25, Textb. 12, top row, extreme right). All the main elements are used by Germanus (cf Knorr 1919, Taf 34, 24-5 for the figure types; Knorr 1952, Taf 28, F for the bush; also Hermet 1934, pl. 100, 19). For the general style see Jacobs 1912, Taf 1, 2. *c* AD 80-110. (10/100, midden beyond Rooms 23-4, possibly of Stage II, II-III elsewhere).
9. (Not illustrated). Form 37. South Gaulish. Four fragments from the same vessel. These show an ovolo with triple tipped tongue, wavy-line borders, a fragmentary figure, stag and leaf motif. *c* AD 80-110. (9/35, demolition of South Range between Stages VI and VII).
10. Form 30. Central Gaulish. Lower part of a decorative panel, somewhat smudged, but including a wavy-line border and opposed double leaves (possibly Rogers', G.365); one border may end in a beaded cup. S&S (pl 39, 460) shows the use of this particular combination of stamps by one of the mould makers supplying Ioenalis (Rogers, X-12) *c* AD 100-20. (0/1, demolition debris post-Stage VIII).
11. (Not illustrated). Form 37. Central Gaulish. The fragment contains a beaded circle used as an ovolo (cf S&S, 35, fig 10, 11; pl 35, 413; pl 40, passim) characteristic of one of the mould makers supplying Ioenalis (Rogers', X-12). *c* AD 100-20. (Topsoil above the bath-house).
12. (Not illustrated). Form 37. Central Gaulish. Rim and a small fragment of ovolo. The ovolo is incomplete but probably the same as Rogers B38 (attributed to his potter X-9, the Medetus-Ranto style of S&S). Trajanic-Hadrianic. (0/80,

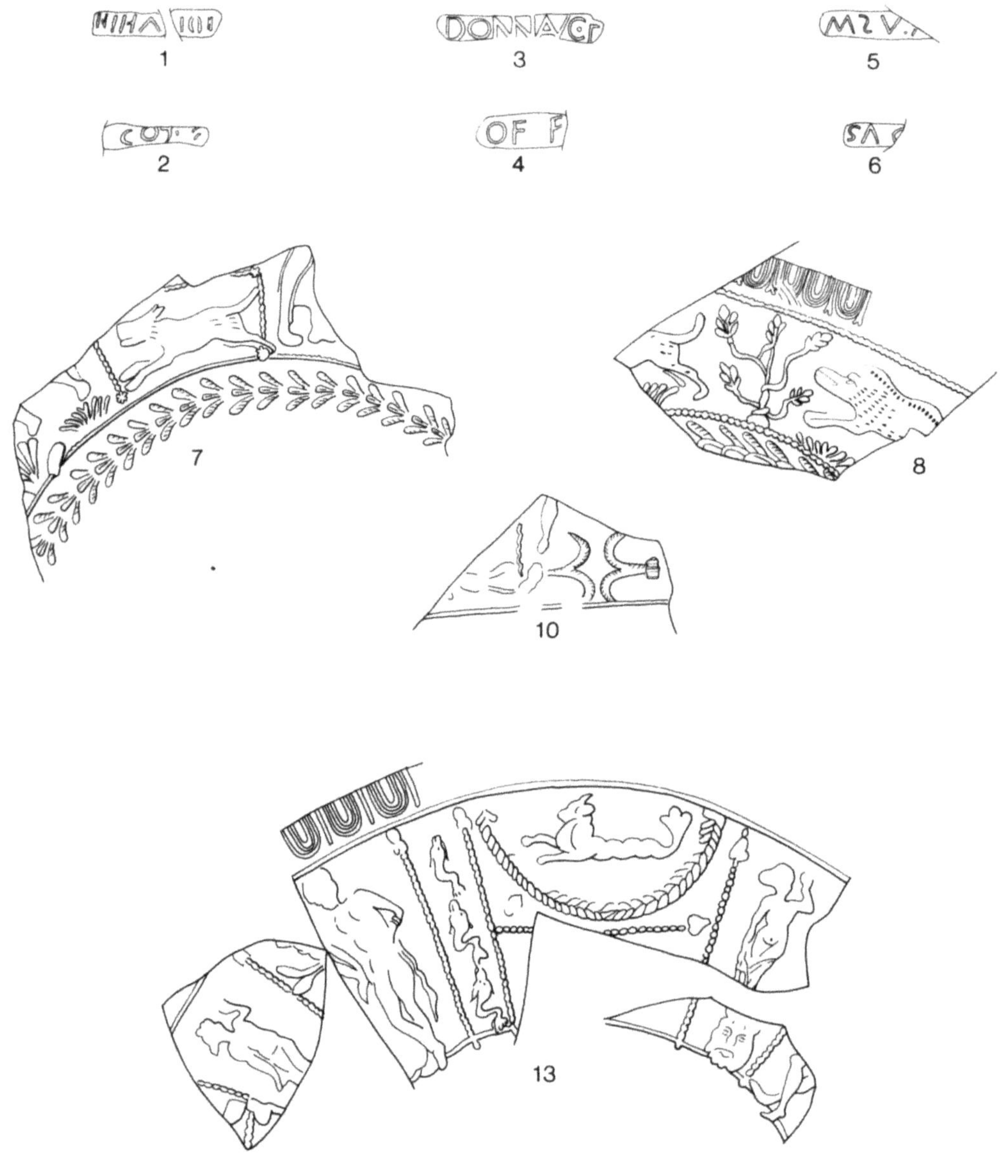

Figure 34: Samian ware, scale 1:2

topsoil west of the complex).

13. Form 37. Central Gaulish. Twelve fragments of the same bowl. The ovolo is Cinnamus ovolo 3a (Simpson and Rogers 1969, fig 1, 1). Below is panel decoration within beaded borders. From left to right the decoration consists of: Venus over a mask, O.310 (apparently used by Divixtus; cf S&S, pl 115, 2, 3); a standing male nude, O.93a (D.56); three dolphins (O.2401) placed nose to tail vertically, so as to resemble a vertical scroll; a sea bull, O.42 within a half-medallion (Rogers', F7, assigned by him to Criciro); and the beginning of a repetition of the sequence. For dolphins used as a 'scroll' cf S&S, pl 155, 22 (Pugnus). The Venus type seems also to have been used by Censorinus and possibly Paternus (cf S&S, pl 102, 11 and pl 109, 7). Oswald attributes his figure type O.93 to Central Gaul and the slightly different O.93a to East Gaul; however, he also equates O.93 with D.56, whereas the Déchelette type is much closer to O.93a. D. 56 is taken from a Lezoux mould. The sea bull and dolphins appear in the work of Cinnamus, which, together with the evidence of the ovolo, makes it likely that this is an example of his work. *c* AD 150-70. (0/11, 0/34, demolition debris above the bath-house and from unstratified levels).
14. (Not illustrated). Form 37. Central Gaulish. The ovolo is incomplete but is probably ovolo 3a of Cinnamus (see no 13 above). Below is an ivy leaf also used by Cinnamus (Rogers, J1) and a bird which also appears in his work (S&S, pl 162, 61). *c* AD 150-70. (Topsoil above the complex).

15. (Not illustrated). Form 37. Central Gaulish. A beaded border with the sea bull O.52a within a medallion. O.52a was used by a number of Antonine potters; of these,Doeccus uses it in combination with the medallion (cf S & S, pl 150, 39, 41). *c* AD 160-90. (Topsoil above the complex).
16. (Not illustrated). Form 72. ? Central Gaulish. A number of fragments of a vessel or vessels decorated in a simple technique between incised lines (cf Oswald and Pryce 1920, pl 77, 4). Probably later second century. From post-Stage VII and unstratified (0/34, demolition debris above ditches west of the complex and from the topsoil above the complex).

Plain samian

17. (Not illustrated). Flagon top. Central Gaulish. Oswald and Pryce 1920, pl LXXXIII, 3; Stanfield 1929, 67-9. (Topsoil above the bath-house).

The coarse pottery (Figs 35-45)
(PVW, DB & KFH, 1976-7)

Pre-complex

18. (Not illustrated). Mortarium flange in orange fabric with a greyish core. The mortarium was possibly made at Wilderspool although the fabric contains very little sandy material. Some grit has been scattered on top of the flange. This vessel was probably made in the early second century. (4/17). (K F H).

Stage I

From Rooms 11:

19. Bowl in orange-buff Cheshire Plain fabric. It has been modelled to resemble a metal bowl with loop handles securing metal rings. For the same motif cf Gillam 1970, type 174; also Hartley and Webster 1973, no 40. Both examples cited are likely to be of the second century. (11/9).

From the partition between Rooms 11 and 11a, sealed by a Stage II floor:

20. Jar in mid-grey fabric; cf Whitchurch (Jones and Webster 1968) no 135 (pre-Hadrianic). (11/73).
21. Jar in grey Cheshire Plain fabric. (11/73). From the same context came a fragment of the the same vessel as no 23. (11/73)

From a layer sealing the partition between 11/11a and 12/12a:

22. Flanged bowl with reeded rim in hard grey fabric; cf Grimes 1930, 85. One of the Flavian-Trajanic series. (11/70).
23. Flanged bowl with reeded rim in light grey Cheshire Plain fabric. Of the same type as no 22. (11/70, with a fragment from 11/73, see above).

Alterations in Stage I

From Room 11:

24. (Not illustrated). Base fragment and shoulder from a lamp of Loeschke type VIII (or less probably Loeschke type V). Almost certainly Italian. Later first century to mid second century AD, but most probably Trajanic-Hadrianic. (11/5). (D B).

Possibly of Stage I

From Yard 8:

25. Flanged bowl in light grey fabric with reeded rim; Corbridge (Richmond and Gillam 1951) no 25 (Trajanic or early Hadrianic) is of the same general type. (7/89).
26. Lid in Black-burnished ware, oxidised pink on the surface. There is internal zig-zag decoration. See Holt (Grimes 1930) no 100 for the form; also Wallace and Webster forthcoming. (7/89).
27. (Not illustrated). Wall sherd of a Black-burnished ware jar with acute-angled lattice decoration. Hadrianic-Antonine. (7/89).

Stage II

From Room 24:

28. Flanged bowl in mid-grey fabric with a darker grey surface. One of the Flavian-Trajanic series; cf Gillam 1970, no 215 for the general type. (10/96).

From Room 73:

29. One of two flanged bowls in light orange/red fabric of Cheshire Plain origin; cf Deanery Field, Chester (Droop and Newstead 1928) pl XLII, 6a. One of the Flavian-Trajanic series. (10/80).
30. Jar in pink Cheshire Plain fabric with grey surface; Wilderspool (Hartley and Webster 1973) no 9 has similarities. (10/74).
31. Jar in pink Cheshire Plain fabric with grey surface. (10/74).
32. Jar in grey Cheshire Plain fabric. The form is reminiscent of Black-burnished jars of Hadrianic-Antonine date. (10/74).
33. Lid in Cheshire Plain fabric varying from pink to grey. (10/74).

From Rooms 24-5:

34. Flanged bowl in grey fabric. One of the Flavian-Trajanic series. (11/51).
35. Flanged dish in Black-burnished ware; cf Gillam (1970) no 307, AD 120-60. (11/51).
36. Flanged dish in Black-burnished ware; cf Gillam (1976) no 61, mid second century. (11/51).

Stage II or III

From midden north of Rooms 24 and 25:

37. Flagon in orange-buff fabric with a cream colour coat. cf Gillam 1970, type 5. AD 110-50. (10/

100).

38. Jar in Black-burnished ware or related fabric. The core has inclusions up to *c* 2 mm in length; cf Gillam 1970, type 120. AD 120-60. (10/100).
39. Jar in light grey fabric. (10/100).
40. Jar in orange-buff fabric, with traces of mica dusting. Perhaps a Holt product. The shape suggests a late first or early second century date. (10/100).
41. (Not illustrated). Wall sherd of jar in off-white fabric; possibly a Holt product. (10/100).
42. Flanged dish in orange-buff fabric, probably of Cheshire Plain origin; cf Holt (Grimes 1930) no 139. Late first or early to mid second century. (10/100).
43. Flanged bowl in Black-burnished ware; probably one of the Hadrianic-Antonine series. (10/100).
44. Flanged bowl in Black-burnished ware. Hadrianic-Antonine; cf Corbridge (Forster and Knowles 1912) no 79. (10/100).
45. Bowl in rose-pink fabric burnt and fired grey in places. The rim appears to be somewhat distorted. (10/100).
46. Lid in orange-buff fabric, probably of Cheshire Plain origin. (10/100).
47. Three fragments from the same mortarium as no 214 below. Manufactured in North Wales or Cheshire. Flavian. (10/100). (K F H).

Stage III

From Rooms 11 and/or 12

48. Two joining pieces forming almost a quarter of a worn mortarium in fine textured, pale brownish cream fabric with white and grey flint trituration grit; a little grit is also scattered over the flange. The form is representative of a type of mortarium known as Bushe-Fox 26-30 (Bushe-Fox 1913, fig 19), found in small numbers all over Britain. They were almost certainly imported. There is considerable variation in the forms, and the significance of these distinctions is not clear. The overall date for the production of this unstamped form is AD 80-120. They are sometimes very tough vessels and could well have remained in use for a long time. (11/6). (K F H).
49. Lid in Black-burnished ware. See no 26 above. (11/38).

From Room 17:

50. Jar in light grey Cheshire Plain fabric. (11/30).

From Alley 7:

51. Jar in Black-burnished ware; cf Gillam 1976, no 2, mid second century. (11/40).
52. Jar in Black-burnished ware; cf Gillam 1976, no 2, mid second century. (11/40)
53. Jar in Black-burnished ware; cf Gillam 1970, no 4, late second century. (11/40).

Stage I, II or III

From the west of the complex:

54. Flanged bowl with reeded rim in light red Cheshire Plain fabric. One of the Flavian-Trajanic series. (4/5).
55. Bowl in light red Cheshire Plain fabric. (4/5).
56. Jar in light brown Cheshire Plain fabric with a darker surface. (12/15).
57. Bowl in light grey Cheshire Plain fabric with darker surface; cf Heronbridge (Hartley and Kaine, 1954) no 59 for a similar vessel but in a different fabric. The Pentre Farm example may be of Wilderspool origin. Probably late first to mid to late second century. (12/15).

Demolition between Stages III and IV

From debris beneath Corridor 33:

58. Jar in light grey fabric. Reminiscent of Hadrianic-Antonine jars in Black-burnished ware and possibly of a similar date. (5/19).
59. Conical cup in off-white fabric with rouletted decoration. A Holt fine ware; cf Greene 1977, fig 58.1, no 2. (Late first to early to mid second century. (7/75). I am grateful to Dr K Greene for examining this piece prior to the appearance of his article.
60. Flanged bowl or dish in Black-burnished ware; cf Gillam 1970, no 307. AD 120-60. (7/80).
61. Jar with lid-seating in fawn-grey fabric with grey surface. (7/88).
62. Tazza in light red fabric. The type was made at both Holt (Grimes 1930, nos 213-4) and Wilderspool (Hartley and Webster 1973, no 59) and this is probably a Cheshire Plain product. (7/88).
63. Mortarium in cream fabric. The type is discussed more fully for no 149 below. Made in Gallia Belgica, perhaps in the Pas de Calais, *c* AD 70-100. (7/88). (K F H).
64. Jar in Black-burnished ware; cf Gillam 1976, no 5. Late second to early third century. (7/88).
65. Bowl in light red fabric with grey core and dark grey smoothed or burnished surface. Possibly a *terra nigra* derivative which has been burnt. For the *terra nigra* type see Greene 1979, fig 47, no 34. (10/92).
66. Dish in light red fabric, perhaps a Holt product; cf Grimes 1930, no 149. Late first to early to mid second century. (10/107).
67. Shallow dish in buff fabric lightly burnished externally and with sooting on the rim; cf Holt (Grimes 1930) nos 132-3. Late first to early to mid second century. (10/107)
68. Jar in Black-burnished ware; perhaps Gillam 1970, no 124. AD 120-60. (11/63).
69. (Not illustrated). Flanged bowl in grey fabric. One of the Flavian-Trajanic series. (10/75).

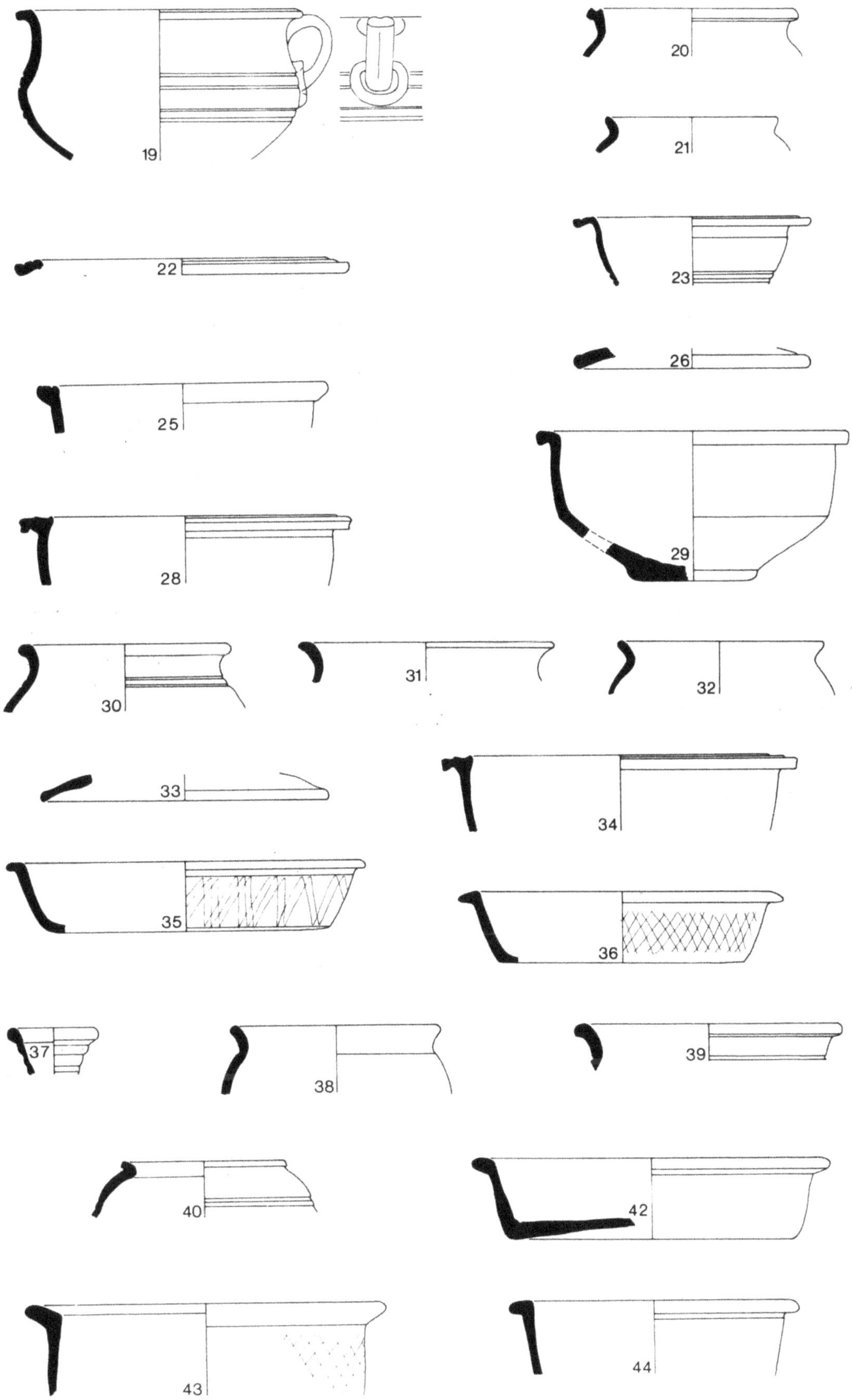

Figure 35: Pottery, scale 1:4

From north of the complex:

70. Jar rim in slightly sandy fabric. (12/22).
71. Flanged bowl with reeded rim in orange fabric; possibly a Holt product, cf Grimes 1930, no 80. Late first to early second century. (12/22).
72. Jar in orange Cheshire Plain fabric. (12/27).
73. Flanged and reeded rim bowl in orange Cheshire Plain fabric; cf Holt (Grimes 1930) no 87. Late first to early second century. (12/27)
74. Flanged bowl with reeded rim in orange-buff Cheshire Plain fabric; cf Holt (Grimes 1930) no 87. Late first to early second century. (12/20).
75. Jar in light orange Cheshire Plain fabric; Holt (Grimes 1930) no 65 is somewhat similar. Late first to early to mid second century. (12/21).
76. Mortarium in fine-textured orange-brown fabric, with tiny traces of what may be a Raetian slip; white quartz trituration grit. This was certainly made in the Cheshire or North Wales area, probably in the early second century. (12/21). (K F H).

Stage IV

From Corridor 33:

77. Flanged dish in Black-burnished ware; cf Gillam 1976, nos 54-5, early to mid second century. (5/10).
78. Dish in Black-burnished ware; Gillam 1976, no 68 (early to mid second century) represents the general type. (5/10).

From north of the complex:

79. Jar in grey Cheshire Plain fabric. (12/24).

From Room 30:

80. Flanged dish in Black-burnished ware; cf Gillam 1976 no 59. Mid second century. (6/42).

From Room 29:

81. Jar rim in grey fabric. (6/58).
82. Lid in grey fabric. (6/58).
83. Jar in light grey Cheshire Plain fabric. (6/61).

From Alley 7:

84. Flagon in orange-buff Cheshire Plain fabric with grey core and traces of a white slip; cf Hartley and Webster 1973, no 3 for the general type. Early to mid second century. (11/15).
85. Flanged bowl in Black-burnished ware; cf Gillam 1976, no 36. Mid second century. (11/15).
86. (Not illustrated). Wall and base of a dish in Black-burnished ware; probably Gillam 1970, no 308. AD 130-80. (11/15).

From Room 18:

87. (Not illustrated). Fragment of cup in orange-buff fabric with traces of mica dusting internally. The vessel appears to be an imitation of the samian form 27; cf Holt (Grimes 1930) no 166; Heronbridge (Hartley and Kaine 1954) no 66. Late first to early to mid second century. (11/19).

Alterations during Stage IV

From Corridor 33:

88. Jar in light grey Cheshire Plain fabric. (5/5).
89. Jar in fairly smooth light grey fabric burnished on the inside of the rim. (5/5).
90. Jar in Black-burnished ware. Possibly Gillam 1976, no 4. Late second century. (5/5).
91. Flanged bowl in light grey fabric. One of the Flavian-Trajanic series. (5/5).
92. Bowl in orange-buff Cheshire fabric with grey core and white slip. (5/5).
93. Lid in dark grey fabric. (5/5).

Demolition between Stages IV and V

94. Jar in Black-burnished ware; cf Gillam 1970, no 128. AD 130-80. (9/40).

Stage V

From Corridor 51A:

95. Jar in light grey ware burnished on the shoulder, and internally on the rim; cf Gillam 1970, no 103. AD 80-120 for the general type. (2/16)
96. Bowl in orange-buff fabric, mica-dusted. A vessel similar to Heronbridge (Hartley and Kaine 1954) no 56 and also of Holt origin seems possible. Late first to early to mid second century. (2/16).

From Room 51:

97. (Not illustrated). Wall fragment of jar with rusticated decoration. Late first to early second century. (2/53).
98. Lid in Cheshire Plain fabric varying from light grey to light brown; cf Gillam 1970, no 339. AD 80-140. (2/53).

From Room 39:

99. Jar in light orange Cheshire Plain fabric with traces of white slip; cf Holt (Grimes 1930) no 64. The shape would be suitable for a 'honeypot'. Late first to early to mid second century. (9/15).
100. Jar in light grey fabric. (9/30).
101. Flanged bowl in light grey fabric. Non-joining fragments from the same level (restored on the drawing) suggest that the bowl was carinated and one of the Flavian-Trajanic series. (9/30).
102. Bowl in light brown fabric. Possibly a vessel reminiscent of samian forms 29 and 30 was intended. (9/30).

From Yard 8:

103. (Not illustrated). Mortarium in light red fabric with white and red trituration grits. Body sherd from a vessel probably made at Wilderspool. (9/44). (K F H).

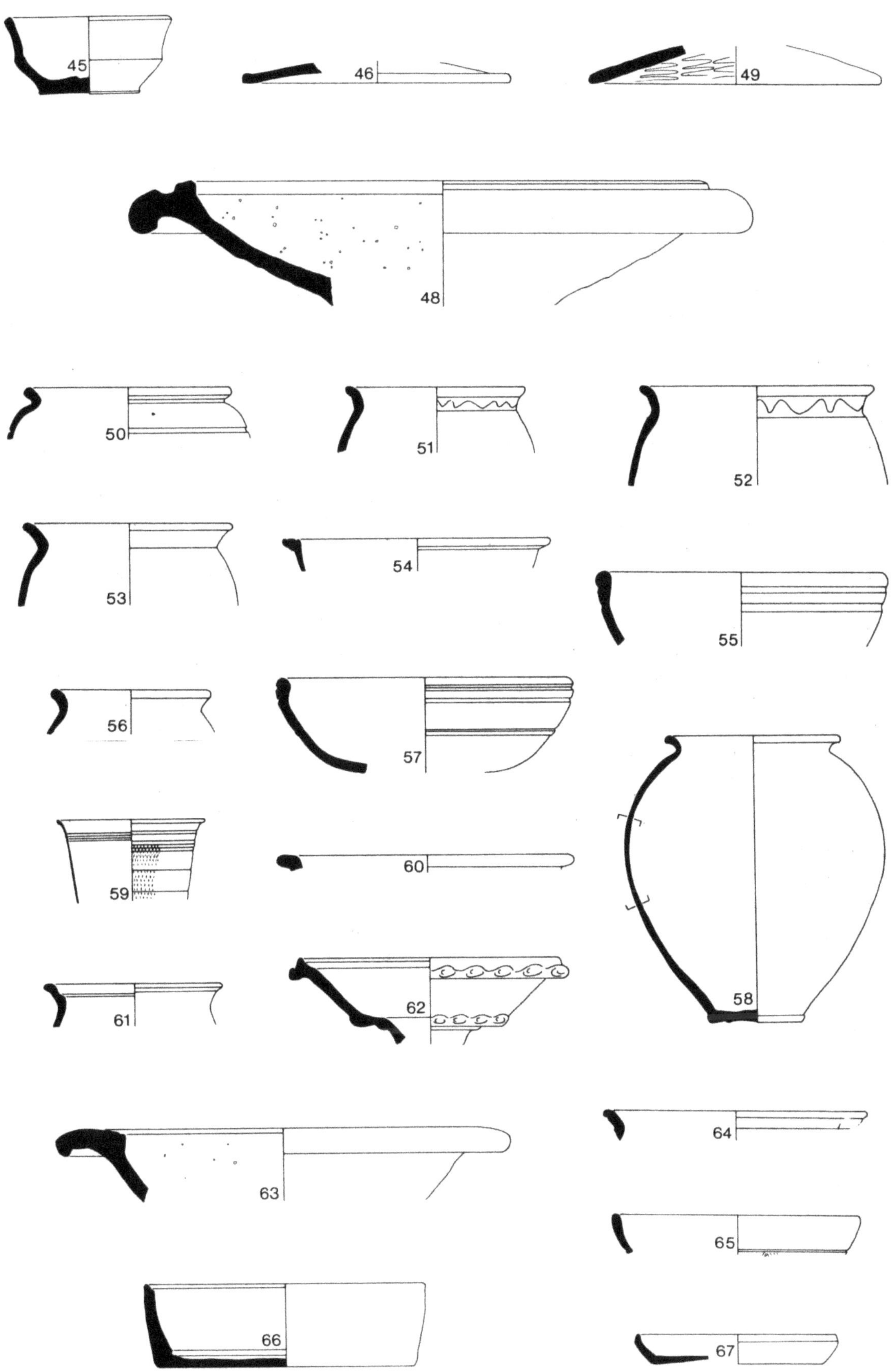

Figure 36: Pottery, scale 1:4

From north of the complex:

104. Bowl in pink-buff fabric; cf Holt (Grimes 1930) no 91. Late first to early to mid second century. (12/32).
105. Lid in mid-grey Cheshire Plain fabric. (12/32).
106. Jar in light grey Cheshire Plain fabric; cf Wilderspool (Hartley and Webster 1973) no 9. Early to mid second century. (12/60).
107. Flanged bowl with reeded rim in light orange fabric; probably a Holt product; cf Grimes 1930, no 83. Late first to early second century. (12/61).
108. Lid in grey Cheshire Plain fabric. (12/61).
109. Dish in Black-burnished ware (BB1). The rim form most closely resembles those of the later half of the second century (eg Gillam 1970 no 310, although that vessel is in BB2). Mumrills (Steer 1960-1) no 18 (mid second century) has similarities. (12/79).
110. Jar in light red-buff Cheshire Plain fabric; cf Holt (Grimes 1930) no 61. Late first to early to mid second century. (12/84).
111. Beaker in pale orange-buff Cheshire Plain fabric with traces of mica dusting externally. The beaker has been decorated with two zones of raised bosses achieved by holding a ring or tube against the outside of the vessel and pushing against it with a finger from inside (finger prints survive on the interior of some bosses to prove this). The rim and fabric resemble Deanery Field, Chester (Droop and Newstead 1928) no 23. The vessel is probably a Holt product. Late first to early second century? (12/86).
112. Jar in Black-burnished ware. The rim appears to be slightly distorted and the angle of the drawing has been arrived at by examination of the shoulder and burnishing marks. Perhaps from a vessel similar to Gillam 1976, no 6 (early third century), although a slightly earlier initial date would be possible. (12/18).

Possibly of Stage V

From the foundations of the dividing wall between Rooms 43 and 44:

113. Jar in Black-burnished ware; cf Gillam 1976, nos 2-3. Mid to late second century. (9/46).

Isolated deposits in the North West Range (Stage V)

From Room 30:

114. Flagon in fine light orange-buff fabric with a white slip; Richmond and Gillam 1955, no 8 (second half of second century) is somewhat similar. (6/18).
115. Jar in Black-burnished ware; cf Gillam 1976, nos 2-3. Mid to late second century. (6/18).
116. Flanged dish in Black-burnished ware; cf Gillam 1976, no 55. Early to mid second century. (6/18).
117. Bowl in orange Cheshire Plain fabric; the form could well have resembled Manchester (Jones and Grealey 1974) no 156, although the fabric is different. Late first to early second century seems likely. (6/18).
118. Portion of a bowl or beaker in thin fine buff fabric, a Holt fine ware; cf Grimes 1930, nos 172 and 191 for possible reconstruction and Greene 1977, for discussion. Late first to early or mid second century. (6/18).
119. Wall sherd in similar ware to no 118. Originally from a vessel such as Holt (Grimes 1930) no 174; cf no 59 above. (6/18).
120. Jar in hard orange fabric, fired grey on the surface externally. Gillam (1970, nos 107-8) illustrates the general type. Late first to early second century. (6/19).
121. Flanged bowl with reeded rim in light orange-buff fabric. The fabric has sparse sandy filler but is probably a Cheshire Plain product; cf Holt (Grimes 1930) no 85. Later first to early second century. (6/19).
122. Flanged bowl or dish in Black-burnished ware. cf Gillam 1976, no 36. Mid second century. Hadrianic-Antonine series. (6/19).
123. Flanged bowl in Black-burnished ware, burnt and partially oxidized. The low chamfer (as Gillam 1976, no 41) suggests a date late in the second century series.
124. Flagon in light red fairly smooth fabric; cf Northwich (Jones *et al.* 1971) no 10 (with references). Late first to early second century. (10/46).
125. Rim probably of a dish in light brown Cheshire Plain fabric with a grey core; cf Holt (Grimes 1930) nos 129-34. Late first to early to mid second century. (10/46).

From Room 29:

126. Small jar or flask in light red Cheshire Plain fabric. Small flasks were being made at Wilderspool (Hartley and Webster 1973, nos 6-8). They appear in the north in the late first and second centuries (cf Gillam 1970, nos 36-9). (6/51).
127. Jar in light brown Cheshire Plain fabric, burnt grey on the rim. (6/51).
128. Jar in light grey fabric with darker surface. A basal sherd from the same layer is apparently from the same vessel and shows lattice decoration, as do sherds from 6/58 (possibly Stage V). An early to mid second century date may be suggested. (6/51).
129. Four fragments, none joining, but probably from the same jar in Black-burnished ware. The angle at the neck varies considerably but the form is probably that represented by Gillam 1970, no 129. AD 140-80. (6/51).
130. Flanged dish in light brown Cheshire Plain fabric. (6/51).

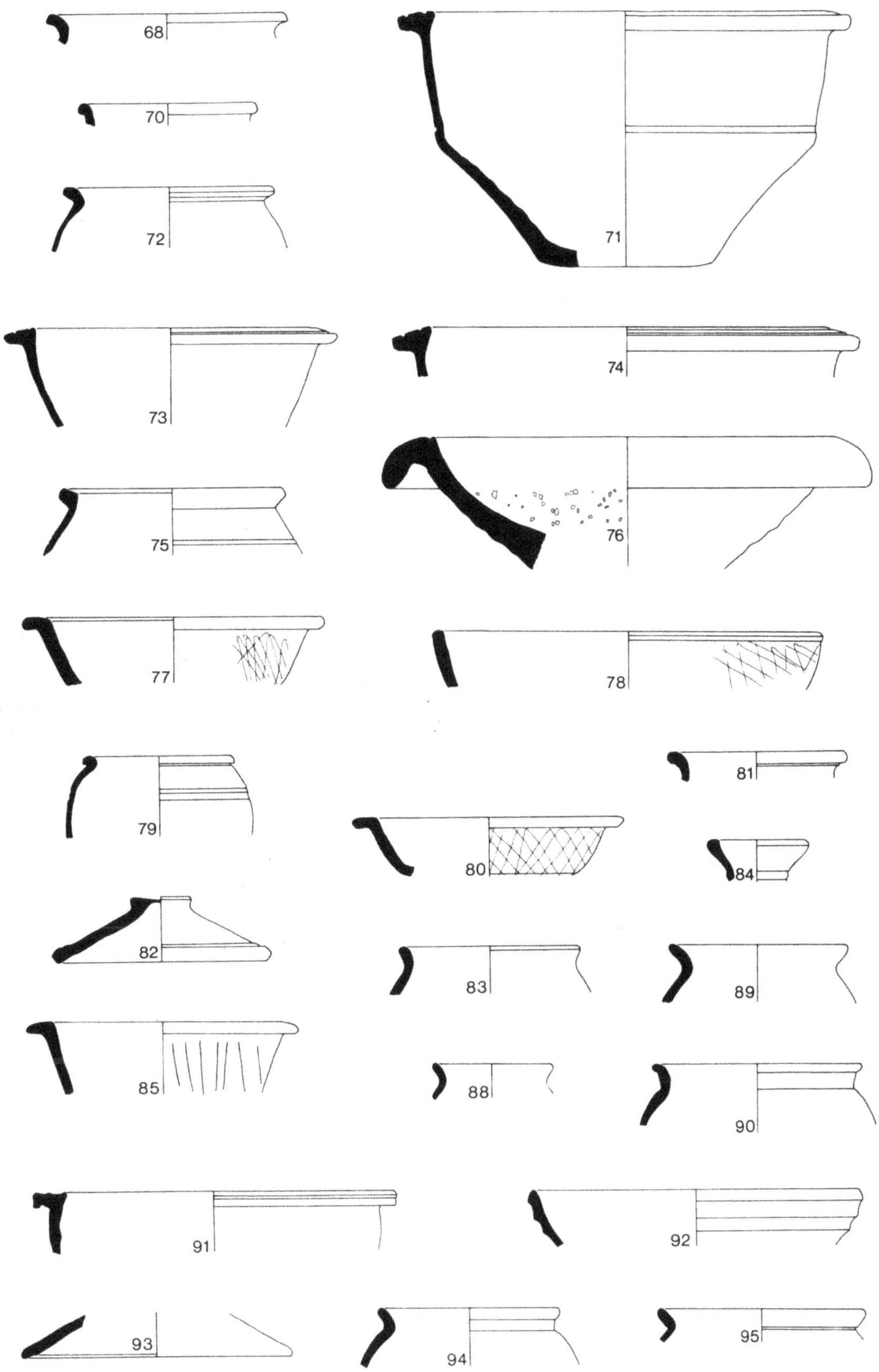

Figure 37: Pottery, scale 1:4

131. Flanged dish in Black-burnished ware; cf Gillam 1976, no 59. Mid to second century. (6/51)

From Room 36:

132. Rim fragment in orange-brown fabric with grey core and dark grey surface. (7/46).
133. Flanged bowl in Black-burnished ware; cf Gillam 1976, no 36. Mid second century. (7/52).
134. (Not illustrated). Twenty-four fragments of a ring-necked flagon broken below the rim. (7/69).
135. Roughcast beaker in white fabric with brown colour coat. Probably a Cologne product; cf Anderson 1981, fig 19.1, no 1 for the rim, although the form must be late as Verulamium (Frere 1972) no 780 (a type dated from *c* AD 150-5 onwards), and Mumrills (Steer 1960) no 79 (mid second century). (7/69).
136. Lid in orange-buff Cheshire Plain fabric burnt grey in places. (7/69).

Stage VI

From Room 7:

137. Flanged bowl with reeded rim in grey Cheshire Plain fabric with darker surface externally and on rim. One of the Flavian-Trajanic series. (1/59).

From Room 51:

138. Dish in Black-burnished ware; cf Gillam 1976, no 77. Late second to early third century. (2/30).

Isolated deposits in the North West Range (Stage VI)

From Room 29:

139. Jar in light grey Cheshire Plain fabric. (6/54).
140. Flanged dish in Black-burnished ware; cf Gillam 1976, no 36. Mid second century. (6/47).
141. Lid in light grey fabric. (6/47).

From Room 30:

142. Beaker in light orange fabric. (6/13).
143. Flanged bowl in light grey Cheshire Plain fabric. One of the Flavian-Trajanic series. (6/13).
144. Flanged bowl in Black-burnished ware; cf Gillam 1976 no 41. Late second century. (6/13).
145. Flanged dish in Black-burnished ware; cf Gillam 1976, no 60. Mid second century. (6/13).
146. (Not illustrated). Very abraded hemispherical bowl in light red-buff Cheshire Plain fabric; reminiscent of samian form 37. (6/13).
147. Dish in Black-burnished ware; cf Gillam 1976, no 52. Mid to late second century. (6/13).
148. Lid in pink to fawn Cheshire Plain fabric burnt grey on the rim. (6/13).

From east of the bath-house:

149. Two joining pieces forming about one quarter of the rim of a mortarium (form: Gillam 1970, no 238) which has been rivetted in at least two places. In fine-textured yellowish cream fabric with flint trituration grit: the grit was carried up to the bead and over the flange. There are traces of concentric scoring inside and on the flange. Mortaria of this type were made in Gallia Belgica, perhaps in the Pas de Calais *c* AD 70-100 (Hartley 1977, Group II). (K F H). (11/64).

From Rooms 32:

150. Flanged dish in Black-burnished ware; cf Gillam 1970, no 309. AD 160-200. (7/34).
151. Jar in orange Cheshire Plain fabric; cf Holt (Grimes 1930) no 59. Late first to early to mid second century. (7/35-6).

Stage VI or later

152. Flanged bowl with reeded rim in grey fabric heavily sooted on the flange. One of the Flavian-Trajanic series. (12/81).

From north of the complex:

153. Flanged bowl with reeded rim in grey Cheshire Plain fabric. One of the Flavian-Trajanic series. (12/8).
154. Bowl or dish in light grey Cheshire Plain fabric. Possibly a vessel reminiscent of the samian form 37 was intended. (12/8).
155. Flagon in light red fabric with orange-buff surface and traces of a white slip; Gillam 1970, no 5 (AD 110-50) represents the general type. (12/12).
156. Possibly a necked bowl in orange-buff Cheshire Plain fabric. (12/12).
157. Bowl in light red Cheshire Plain fabric. Reminiscent of the samian form 37; cf Heronbridge (Hartley and Kaine 1954) no 58; Holt (Grimes 1930) nos 153-4. Late first to early to mid second century. (12/12).
158. Flagon in orange-buff Cheshire Plain fabric with traces of a white slip; cf Corbridge (Richmond and Gillam 1955) no 8 (late second century); Milecastle 48 (Gibson and Simpson 1911) no 14, (from Period II); Birrens (Robertson 1975) fig 75, 3. A mid or late second to mid third century date may be suggested for the type. (12/13).
159. Jar in Black-burnished ware; cf Gillam 1970, nos 123-5. AD 120-80. (12/13).
160. Bowl in light brown Cheshire Plain fabric. Reminiscent of the samian form 37. (12/13).
161. (Not illustrated). Wall fragments of a roughcast jar in orange-buff Cheshire Plain fabric; cf Wilderspool (Hartley and Webster 1973) nos 23-33. Early to mid second century. (12/13).
162. Jar in light grey Cheshire Plain fabric. (12/14).
163. Flanged bowl in Black-burnished ware; cf Gillam 1976, nos 35-6. Mid second century. The vessel has been repaired with a rivet. (Joining fragments from 12/13 and 12/14).
164. Bowl or dish in buff Cheshire Plain fabric. (12/14).

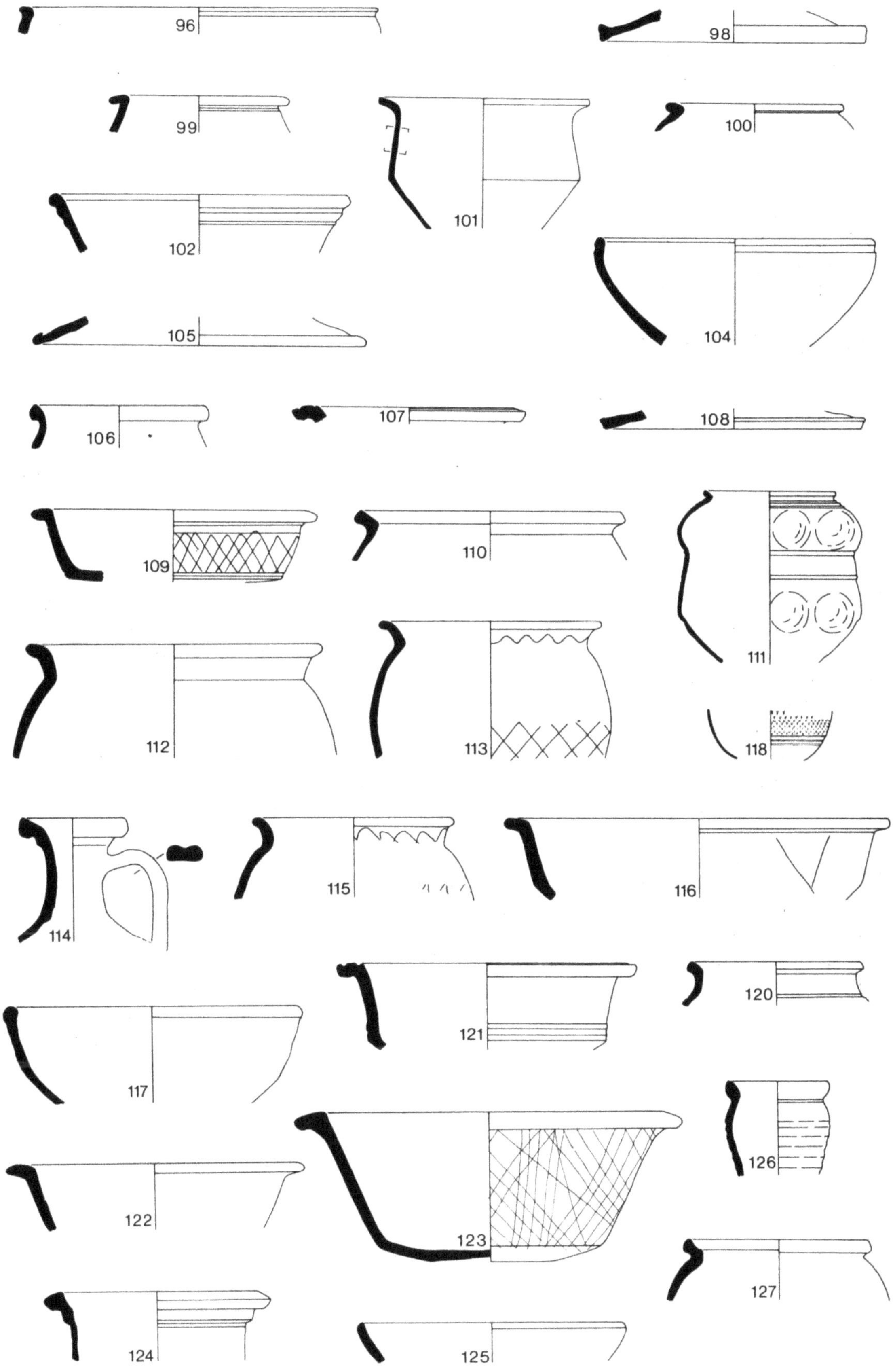

Figure 38: Pottery, scale 1:4

165. Lid in pink Cheshire Plain fabric with a grey surface. (12/14).
166. Flanged bowl in grey Cheshire Plain fabric with a darker surface. Presumably one of the Flavian-Trajanic forms but perhaps late in the series. ?Early second century. (12/11).
167. Flanged bowl in light red Cheshire Plain fabric; cf Holt (Grimes 1930) no 139. ?Early second century. (12/11). Layer 12/11 also contained a sherd of Black-burnished ware and part of a Wilderspool roughcast beaker.
168. (Not illustrated). Bowl in red fabric with grey core and white slip. Almost certainly part of the same vessel as no 92. (2/20).
169. Bowl in light red fabric with applied handle. The form resembles one common in the Severn Valley; cf Webster 1976, nos 145-53, and, although exact parallels are lacking, this is probably a Severn Valley product. (2/20).
170. (Not illustrated). Mortarium base, burnt. Probably a Wilderspool product. (4/3). (K F H).

Demolition between Stages VI and VII

From the destruction levels in Rooms 11-12:

171. (Not illustrated). Body sherd of a bowl in light orange-red fabric with patchy red colour coat. A Holt product. Joins no 184. Late first to early to mid second century. (8/37).
172. Beaker in brown Cheshire Plain fabric with darker surface. Probably a Wilderspool product; cf Hartley and Webster 1973, nos 31-33. Early mid second century. (8/27).
173. (Not illustrated). Handle fragment from a lamp of Loeschke type VIII. Gaulish. Late first century to mid second century. (8/32). (D M B).
174. Jar in orange-buff Cheshire Plain fabric. (8/32).
175. Rim in grey Cheshire Plain fabric. Possibly part of a wide-mouthed flagon. (8/32).
176. Jar in Black-burnished ware; cf Gillam 1976, no 3. Late second century. (11/3).

From Room 43:

177. Jar in grey Cheshire Plain fabric. (9/35).
178. Flanged bowl or dish in Black-burnished ware. One of the Hadrianic-Antonine series. (9/35).

From Room 44:

179. Lid in Black-burnished ware; cf Holt (Grimes 1930) no 100, for a complete example of the type. Also Wallace and Webster forthcoming. (1/44).

Stage VII

From Room 51:

180. Beaker in smooth orange-brown fabric. (2/15).
181. Wide-mouthed jar in orange-buff Cheshire Plain fabric. Despite the likely origin of the fabric the form is related to vessels found most frequently in Severn Valley ware; cf Webster 1976, nos 20-31. Similar jars are found at Wilderspool on the site excavated in 1976 and occupied into the late second or early third century (Williams and Hinchcliffe, forthcoming). (2/15).
182. Mortar-like bowl in light orange-buff fabric with grey core. This is probably a Cheshire Plain product although the fabric contains relatively little sandy filler. The form is presumably related to that of contemporary mortaria and is most likely, therefore, to be of second century date. (2/15).

From Room 42:

183. Flanged bowl with reeded rim in grey Cheshire Plain fabric with darker surface. One of the Flavian-Trajanic series. Mortar is adhering to parts of the vessel. (8/28).
184. Bowl in light orange-red fabric with red colour coat. A vessel reminiscent of the samian form 37 was probably intended. This is a Holt product. I am grateful to Dr P Carrington for commenting on the piece. Cf Greene 1977, no 117 for some discussion of Holt red-slipped fine wares. Late first to early second century. This vessel consists of several joining fragments; from Room 42 (8/24, two fragments); Room 45 (8/30), one fragment); from the destruction levels in Rooms 11 and 12 prior to Phase VII and a further fragment (8/37), cf no 171.

From Yard 46:

185. Flanged bowl or dish in orange to light brown fabric. Perhaps one of the Flavian-Trajanic series. (1/65).

From Room 45:

186. Bowl in grey Cheshire Plain fabric. (8/12).

From Room 40:

187. Roughcast beaker in light brown Cheshire Plain fabric; cf Wilderspool (Hartley and Webster 1973) no 24. Early mid second century. (9/17).
188. Mortarium in coarse light red fabric with traces of a cream slip. Made in Wales or in the Cheshire Plain. (9/17). (K F H).

From Corridor 51A:

189. Beaker in smooth light grey fabric. Burnished above the decoration. (2/21).
190. Rim fragment of a flanged and grooved bowl in Black-burnished ware; cf Gillam 1970, no 226. AD 200-240. (2/12).

Isolated deposits in the North West Range (Stage VII)

From Corridor 33:

191. Beaker in Black-burnished ware burnt from grey to orange externally; cf Gillam 1976, no 29. Mid to late second century. (5/4).
192. Dish in Black-burnished ware. The fabric is BB1

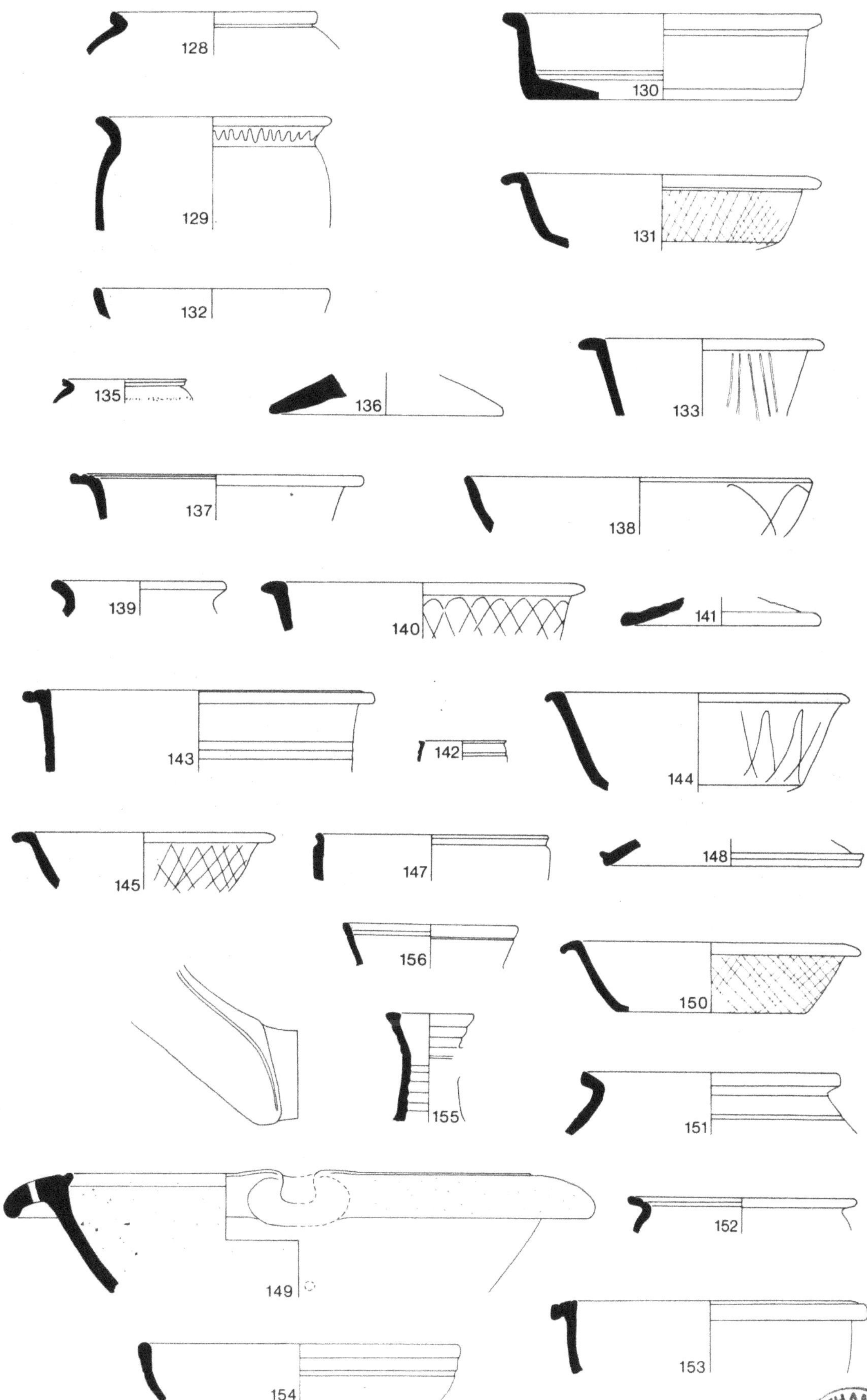

Figure 39: Pottery, scale 1:4

but the form most closely resembles that of BB2, eg Gillam 1970, no 310, AD 150-210; Steer 1960-1, no 48, (mid second century). A date in the second half of the second century seems the most likely. (5/4)

193. Roughcast beaker in light orange fabric with a dark brown colour coat; North Gaulish cf Anderson 1981, fig 19.3, no 29. Mid second century. (7/31-2).
194. Lid in pink-buff Cheshire Plain fabric. (7/31-2).

From Room 29:

195. Beaker in orange-brown Cheshire Plain fabric; cf Wilderspool (Hartley and Webster 1973) no 23. Early mid second century. (6/46).
196. Jar in light grey Cheshire Plain fabric. (6/46).
197. Jar or possibly large beaker in Black-burnished ware; Gillam 1976, nos 30-1, represent the general type. Early to mid second century. (6/46).
198. Flanged dish in Black-burnished ware; cf Gillam 1976, no 60. Mid second century. (6/46).
199. Dish in Black-burnished ware. (6/46).

From Room 30:

200. Jar or flagon neck in orange-buff Cheshire Plain fabric with a white slip. (6/12).

From Room 36:

201. (Not illustrated). Wall fragment of beaker in buff with brown colour coat. Probably a Cologne product. Possibly from a vessel such as Anderson 1981, fig 19.1, no 5 or 6 (late first to late second centuries). But see also Corder 1961, fig 18, nos 3-4. Almost certainly later than *c* AD 160. See no 229 below. (7/16).
202. Dish in light grey Cheshire Plain fabric. (7/16).
203. Dish in orange-buff Cheshire Plain fabric, burnt on rim. (7/16).
204. Very abraded bowl in light grey Cheshire Plain fabric. (7/18).
205. (Not illustrated). Part of the lower wall of a jar in pink Cheshire Plain fabric with decoration of scored diagonal lines. The six lines *c* 1 mm wide and 1 mm apart were probably scored with a comb. (7/19).
206. Jar in smooth grey fabric. (7/23).
207. Bowl in pink-buff Cheshire Plain fabric. Probably one of the series represented by Holt (Grimes 1930) nos 129-37. Late first to early to mid second century. (7/23).

From Room 29:

208. Jar in grey fabric. (6/35).
209. Jar in light grey Cheshire Plain fabric. (6/35).
210. Jar in buff fabric probably of Cheshire Plain origin. There is an oxidized deposit on part of the rim (? lead); cf Gillam 1970, no 117. AD 130-50. (6/35 and 6/36).
211. Lid in grey Cheshire Plain Fabric. (6/36).
212. Lid in Black-burnished ware; cf no 179 above. (6/36).
213. Flanged bowl in Black-burnished ware; cf Gillam 1976, no 34. Early to mid second century. (6/35 and 6/36).
214. Two joining pieces making up about a quarter of a large hook-rimmed mortarium in purplish pink fabric with blue-grey core. The surface is fired to orange-brown and there are traces of a cream slip. There is abundant white quartz trituration grit. The fabric contains a lot of sandy material. The very deeply hooked rim undoubtedly points to a Flavian date. Manufacture in North Wales is probable. Joins with no 47. (6/36 and 80). (K F H).
215. Jar in Black-burnished ware; cf Gillam 1976, no 3. Late second century. (6/48).

From Room 31:

216. Lid in Black-burnished ware, cf no 179 above. (7/103).

Demolition Levels between Stages VII and VIII (sealed only by post-medieval or modern layers).

From post-Stage VII layers:

217. Jar in light red Cheshire Plain Fabric. (0/38).
218. (Not illustrated). Lid in light red Cheshire Plain Fabric. (0/38).
219. (Not illustrated). Lid as no 216. (0/39).
220. Lid in grey Cheshire Plain Fabric; cf Gillam 1970, no 339. AD 80-140. (0/8).
221. (Not illustrated). Handle from a lamp of Loeschke type VIII (or less probably Loeschke type V). Almost certainly Italian. Late first to mid second century AD and most likely Trajanic-Hadrianic. (0/12). (D M B).
222. (Not illustrated). Two fragments from different roughcast beakers in Cologne fabric (white fabric with grey-brown colour coat); cf nos 201 an 229. (0/12).
223. Jar in pink-buff Cheshire Plain fabric. A loop handle resembling that used on a honey pot has been affixed 'sideways on'. It could not have been used to tie on a lid but must have aided lifting. (0/12).
224. Jar in Black-burnished ware sooted on rim; cf Gillam 1970, nos 132-134. Mid second to early third centuries. (0/12).
225. Bowl in grey Cheshire Plain fabric. There is some resemblance in form to the green-glazed hemispherical bowls probably produced at Holt; cf Whitchurch (Jones and Webster 1968) nos 145 and 145a. (0/12).
226. Hammerhead mortarium in off-white fabric, burnt, probably after fracture. A reeded hammerhead mortarium made in the Mancetter-Hartshill potteries *c* AD 230-80. A small body sherd from a Mancetter-Hartshill mortarium comes from

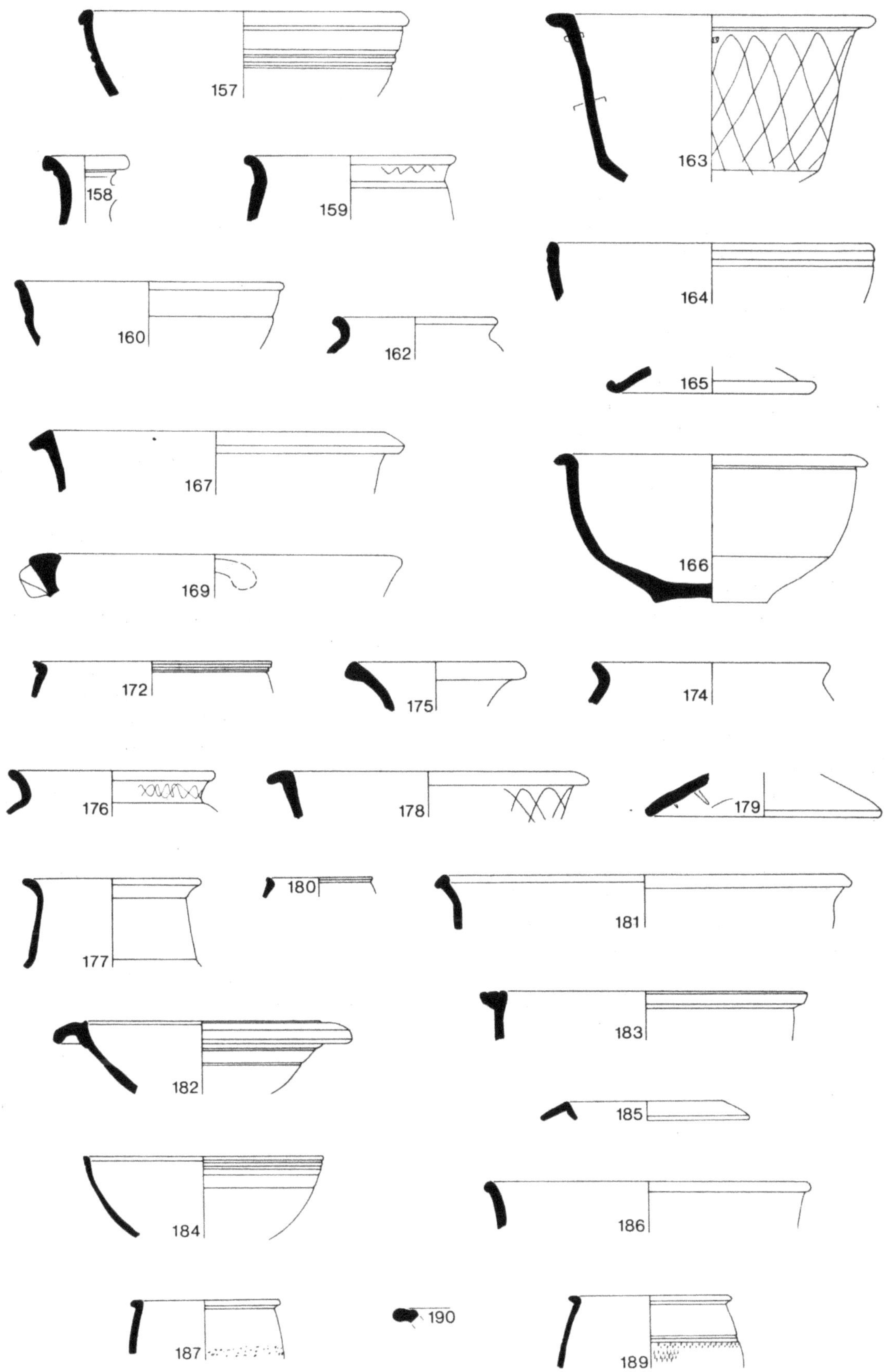

Figure 40: Pottery, scale 1:4

the same layer. (0/12). (K F H).

227. Amphora stopper in light buff fabric; cf Atkinson 1942, no A14 for an illustration of an example found *in situ* in the neck of a globular amphora. Examples from the Neronian fortress at Usk (eg from pit DTZ, drawing 3530 in the Usk archive) demonstrate that this type of stopper was in use from an early date. See also Colls *et al*, 1977, 24 and 39. (0.12).
228. Lid in grey Cheshire Plain fabric. (0/12).
229. Beaker in light pink fabric with brown colour coat and roughcast decoration. Probably a Cologne product, see Anderson 1981, fig 19.1, no 6 (mid to late second century) also Frere 1972, no 781. Mid second century. (0/26).
230. Reeded hammerhead mortarium in off-white fabric made in the Mancetter-Hartshill potteries, probably later than AD 260 and possibly in the fourth century. (0/26). (K F H).
231. Rim of jar in Black-burnished ware. Possibly part of a late jar such as Gillam 1976, nos 12-14 (fourth century). (0/35).
232. (Not illustrated). Flange and part of wall of a flanged and ridged bowl in Black-burnished ware; Gillam 1970, 228. AD 290-370. (0/35).
233. Jar in Black-burnished ware; possibly Gillam 1976, no 10. Late third century. (0/36).
234. Bowl in grey Cheshire Plain fabric; cf Wilderspool (Williams and Hinchcliffe forthcoming); Holt (Grimes 1930) no 92, second century (0/36).
235. Tazza in light orange-buff fabric with a white slip; cf Holt (Grimes 1930) nos 213-4; Wilderspool (Hartley and Webster 1973) nos 59-60. (0/36 and 0/1).
236. Mortarium in typical Wilderspool fabric with cream slip; similar to Hartley and Webster 1973, fig 11, no 91, though the flange of this example is shallower. Primarily Hadrianic. (0/56). (K F H).
237. Three pieces from the one partly burnt mortarium in orange-brown fabric containing some sandy particles, with traces of cream slip and with quartz trituration grit. The fabric is not typical of Wilderspool but mortaria of this type and fabric are undoubtedly common in North Wales and Cheshire. Production at Holt is possible but there were probably many small workshops in this general area in the early second century. (0/56). (K F H).
238. (Not illustrated). Fragments of mortarium in white Mancetter-Hartshill fabric, AD 140-400. (0/20). (K F H).
239. (Not illustrated). Fragments of mortarium in white fabric made in the Mancetter-Hartshill potteries, AD 140-400. The level also contained a body sherd, probably from the same vessel as no 307. (0/6). (K F H).
240. Dish in Black-burnished ware; typologically it is intermediate between Gillam 1970, no 316 (AD 125-60) and no 329 (AD 190-340). There are similarities to Gillam 1976, no 79. A date within the second half of the second or early third century may be suggested. (0/6).
241. (Not illustrated). Wall fragment of a beaker in off-white fabric with maroon colour coat externally and light brown colour coat internally. Apparently from a vessel such as Gillam 1970, no 86. From Cologne or the Nene Valley. Later second or early third centuries. AD 180-230). (0/3).
242. Rim in smooth light red fabric with traces of red colour coat. Possibly an imitation of the samian form 38. See also Hartley 1981, for red slip wares from Wilderspool. Late second century or later. (0/2).
243. A flange fragment from a slightly burnt mortarium in slightly sandy, cream fabric with pink core. The fabric and the sharply turned-under flange is typical of the mortaria produced in workshops like those at Cowley, Headington and elsewhere in the vicinity of Oxford. It approximates to Young 1977, M17 which Young dates to *c* AD 240-300. (0/2). (K F H).
244. Beaker in light red fairly smooth fabric. (0/1).
245. Beaker in Black-burnished ware; cf Gillam 1976, no 66. Early to mid second century. (0/1).
246. Jar in mid to dark grey fabric containing some sandy filler and possibly of Cheshire Plain origin. The form is intermediate between the everted rim jars of the late first or early second century and the latticed jars in Black-burnished ware of the Hadrianic-Antonine period. Gillam 1970, no 117 (AD 130-50) has similarities. (0/1).
247. Jar in Black-burnished ware; Gillam 1970, no 118, (AD 120-60) represents the general type. (0/1).
248. Jar in Black-burnished ware. (0/1).
249. Jar in Black-burnished ware; Gillam 1976, no 16. Early to mid second century. (0/1).
250. Jar in Black-burnished ware; cf Gillam 1976, no 10. (0/1).
251. Jar in Black-burnished ware; three fragments probably from the same vessel; cf Corbridge (Richmond and Gillam 1951) nos 29 and 41 (late third to fourth century). (0/1).
252. Wide-mouthed jar in smooth light pink-buff fabric with little filler; probably Severn Valley ware; cf Whitchurch (Jones and Webster 1968) nos 177 and 177a, late second century. (0/1).
253. Flanged and grooved bowl in Black-burnished ware; cf Gillam 1976, no 227. AD 210-300. (0/1).
254. Flanged dish in Black-burnished ware; one of the Hadrianic-Antonine series. (0/1)
255. Dish in Black-burnished ware; Gillam 1976, no 77. Late second to early third century. (0/1).
256. Raetian mortarium with much quartz trituration grit. The rim-profile is similar to that of mortaria with cream slip made at Wilderspool and stamped

Figure 41: Pottery, scale 1:4

by potters using trademarks (Hartley and Webster 1973, fig 11, 95). This vessel was made either in Cheshire or in North Wales *c* AD 100-35. (0/1). (K F H)

257. Flange fragment from a Raetian or sub-Raetian mortarium with distal bead, probably made at Wilderspool. These mortaria cannot be dated closely but this example would fit a date in the mid second century. (0/1). (K F H).

Stage VIII

From Room 49:

258. Small jar or beaker in Black-burnished ware; (2/17).

From Room 52/53:

259. Jar in fawn to light brown Cheshire Plain fabric; cf Holt (Grimes 1930) no 60. Late first to early or mid second century. (3/8).
260. (Not illustrated). Fragment of indented beaker in white fabric with dark grey-brown colour coat. Probably from the Nene Valley. Although this type of beaker is more common from the mid third century onwards (cf Gillam 1970, no 54) some examples do occur from mid to late second century (cf Frere 1972, no 790). (3/37).

Stage VII or VIII

261. Dish in Black-burnished ware; cf Gillam 1976, nos 68-9. Early to mid second century. (4/1).
262. Lid in grey Cheshire Plain fabric. (4/1).
263. Flagon in pink to pink-buff Cheshire Plain fabric with traces of white slip; cf Wilderspool (Hartley and Webster 1973) no 3; Turret 48B (Shaw 1926) no 16 (Period Ib). ? Mid to late second century. (0/42).
264. Rim in pink Cheshire Plain fabric with a light orange-buff surface. Possibly part of a large flagon; cf Corbridge (Richmond and Gillam 1955) no 8. Late second century. (0/42).
265. Handled beaker in Black-burnished ware; cf Gillam 1976, no 25. Mid second century. (0/42).
266. Flanged dish in Black-burnished ware; cf Gillam 1976, no 65. Late second century. One of two Hadrianic-Antonine flanged dishes. (0/42).
267. Flanged and grooved bowl or dish in Black-burnished ware; cf Gillam 1976, no 42. Late second to early third century. (0/42).
268. Flanged bowl with broken flange in light red fabric. Perhaps intended to be reminiscent of the samian form 38 and, if so, then late second century or later. (0/42).
269. (Not illustrated). Mortarium in orange-buff fabric. A body sherd from a vessel made at Wilderspool or in the vicinity. Second century. (0/42). (K F H).
270. (Not illustrated). Dish in Black-burnished ware. Very similar to no 261. (0/42).

Stage VIII or later (sealed only by post-medieval or modern layers)

From the sequence west of the complex:

271. Flagon in light orange fabric with traces of a white slip. (0/44).
272. Jar in Black-burnished ware; cf Gillam 1976, nos 7 and 10. Third century (0/44).
273. (Not illustrated). Sherd from a flanged and ridged dish in Black-burnished ware. The piece has been burnt and the flange broken off. Gillam 1970, no 228 (AD 290-370) represents the general type. (0/44).
274. (Not illustrated). Wall fragment of beaker in Nene Valley fabric (white fabric with dark grey-brown slip). Fragment of trailed decoration as on Gillam 1970, no 51. AD 250-300. (0/45).
275. (Not illustrated). Mortarium spout in light red fabric made at Wilderspool. Of the same basic type as no 280 but with traces of red-brown slip used on the Raetian and sub-Raetian mortaria made there. *c* AD 100-30. (0/71). (K F H).
276. Mortarium in cream fabric. This rim profile was produced in the Mancetter-Hartshill potteries *c* AD 140-240. Very large mortaria like this example were rarely made in these potteries during the period when mortaria were stamped there, so the date of this piece is more likely to be post *c* AD 180. (0/71). (K F H).

Alterations in Stage VIII

From Yard 16:

277. A mortarium with nearly vertical collar in hard pink-brown fabric fired to drab cream at the surface and tempered with tiny quartz grit. The trituration grit, also quartz, is continued up to the bead and some is embedded in the outside of the collar. An import from the Rhineland. Such mortaria were made in a known workshop at Soller, Kr Duren in Lower Germany and no doubt in many others from *c* AD 150 probably until the mid third century. Mortaria of this type have been found at Mumrills, where they are to be dated within the period AD 140-65 and in a late second century deposit at Clausentum. (1/9). (K F H).
278. Flanged bowl in Black-burnished ware; one of the Hadrianic-Antonine series. (1/10).
279. (Not illustrated). Mortarium in white fabric, body sherd. Mancetter-Hartshill potteries. AD 140-400. (1/10). (K F H).
280. Mortarium in orange-buff fabric. The form and fabric of this beadless mortarium are typical of many made in the kilns at Stockton Heath, Wilderspool. This example has no trace of any surface slip but this form usually had a red-brown Raetian-type slip on top of the flange and bead, which is reminiscent of samian ware. There is a clear similarity to mortaria made at Holt

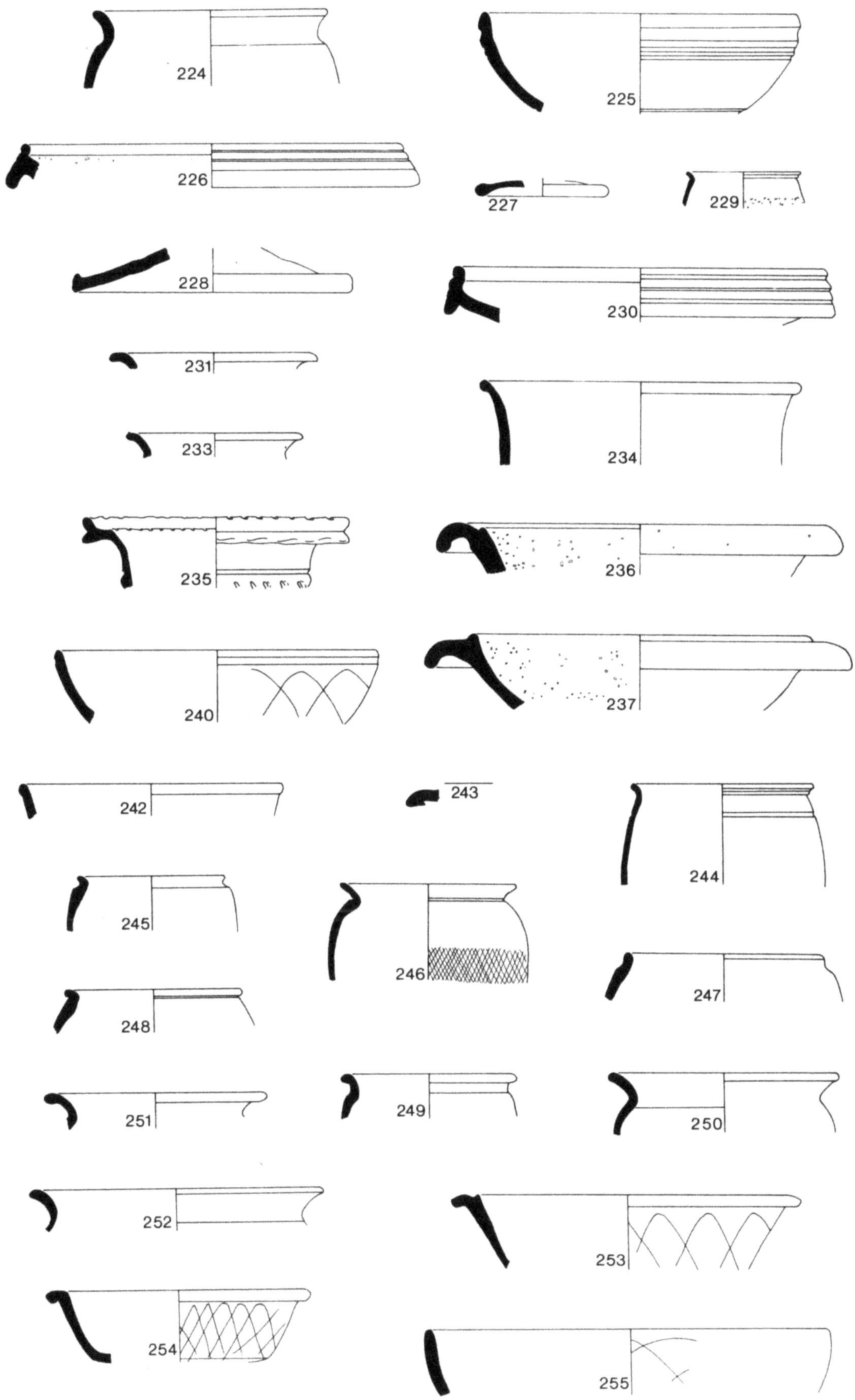

Figure 42: Pottery, scale 1:4

(Grimes 1930, fig 61, nos 2, 3, 5) and a date in the the first half of the second century is virtually certain. (1/15). (K F H).

The northern part of the site

281. Mortarium in orange-buff, Wilderspool fabric (no slip surviving). The vessel approximates to Hartley and Webster 1973, fig 11, 89. It was probably made in the Hadrianic-Antonine period. (13/23). (K F H).
282. Very abraded jar in orange Cheshire Plain fabric. The surface has been completely lost, removing much of the decoration which was rouletted. For the rim form, cf Hartley and Webster 1973, nos 24-5 and for the decoration, (ibid) nos 35-7. Early to mid second century. (13/22).
283. Flanged bowl in Black-burnished ware; cf Gillam 1976, no 38. Mid to late second century. (13/22).
284. Mortarium in orange-buff with grey core (no slip surviving). A beadless mortarium in typical Wilderspool fabric, probably made in the mid second century. (13/17). (K F H).
285. Flanged dish in Black-burnished ware burnt light grey; cf Gillam (1976) no 56. Early to mid second century. (13/9).
286. 'Honey pot' in orange-buff Cheshire Plain fabric with traces of a white slip externally; Holt (Grimes 1930) no 70, has similarities. (13/9).
287. Lid in orange-buff Cheshire Plain fabric. (13/8).
288. Jar in orange-buff Cheshire Plain fabric; possibly part of a 'honey pot'. (13/2).
289. Jar in grey Cheshire Plain fabric. (13/2).
290. Jar in fawn-grey fabric. (13/2).
291. Lid in Black-burnished ware. (13/2).

292-3. (Not illustrated). Two examples of flanged bowls in Cheshire Plain fabric (one grey, the other orange-buff). Both are probably Holt products (cf Grimes 1930, no 84) and belong to the Flavian-Trajanic series. (13/2).

Post-complex levels

294. Jar in Black-burnished ware; possibly Gillam 1976, no 4. Late second century. (6/21).
295. Rim of jar in Black-burnished ware. Only a short length of rim survives and none of the wall of this vessel but it apparently belongs to a vessel such as Gillam 1970, no 146. AD 280-350. (6/21).
296. Bowl in light orange-red Cheshire Plain fabric. (6/21).
297. (Not illustrated). Two flanged bowls in Black-burnished ware. Hadrianic-Antonine. (6/21).
298. (Not illustrated). Part of the base of a wide-mouthed jar in Severn Valley ware. (6/21).

Ploughsoil (unstratified)

Among the quantity of Roman pottery from the post Roman soil accumulation were the following (all are from topsoil above the complex):

299. Flagon in light red fabric, probably of Cheshire Plain origin, slightly reduced on the surface; Northwich (Jones *et al* 1971) fig 11, no 15; fig 16, no 10, has certain similarities. ? Late first to early second century.
300. Flagon in light red Cheshire Plain fabric. The rings on the neck are evenly placed, a feature generally associated with an early date (cf Gillam 1970, nos 2, 4). See also Wilderspool (Hartley and Webster 1973) nos 2 (probably not earlier than *c* AD 90). A late first century date seems probable.
301. Flagon in light red fabric with traces of white slip; Gillam (1970) no 7 (AD 130-220) is similar.
302. (Not illustrated). Rim of beaker in light brown fabric with dark grey-brown colour coat. Only a fragment of rim survives but the original beaker probably resembled Gillam 1970, no 46. AD 220-60.
303. (Not illustrated). Base of a beaker in very light orange-buff fabric with orange colour coat. Possibly a Lezoux product. The base is very abraded but would suit a vessel such as Gillam 1970, no 48 (early mid third century); cf Brewster 1972, no 4, and Greene 1978a, 18-19 for further discussion.
304. (Not illustrated). Body sherd of beaker in light grey fabric with a grey core and dark grey-brown colour coat. Probably part of the shoulder of a beaker such as Gillam 1970, no 44. AD 190-220 and a Trier product. See Brewster 1972, for a discussion; Greene 1978a, 18-19; Greene 1978b, 5.
305. Jar in coarse light red Cheshire Plain fabric.
306. Fragment of bowl in thin grey fabric with a patchy olive green glaze both inside and out. Probably part of the base of a bowl such as Grimes 1930, fig 76,a. For a discussion of this class of Holt fine ware, cf Greene 1977, 122, fig 8.1.
307. Jar in Black-burnished ware; cf Gillam 1976, no 8. Mid third century.
308. Jar in Black-burnished ware; possibly Gillam 1976, no 10. Late third century.
309. Jar in Black-burnished ware; cf Gillam 1970, no 134 (AD 160-230).
310. One of at least two jars in Black-burnished ware; cf Gillam 1976, no 12. Early fourth century.
311. Wide-mouthed jar in orange-buff Severn Valley ware; cf Webster 1976, no 26 (? late second to late third century).
312. Flanged bowl in orange-buff to grey-brown Cheshire Plain fabric. One of the Flavian-Trajanic series.
313. Mortarium in orange-buff sandy fabric with a grey core. A Raetian-type mortarium with traces of the red-brown slip normal on the flange of such mortaria. Probably made in the Wilderspool

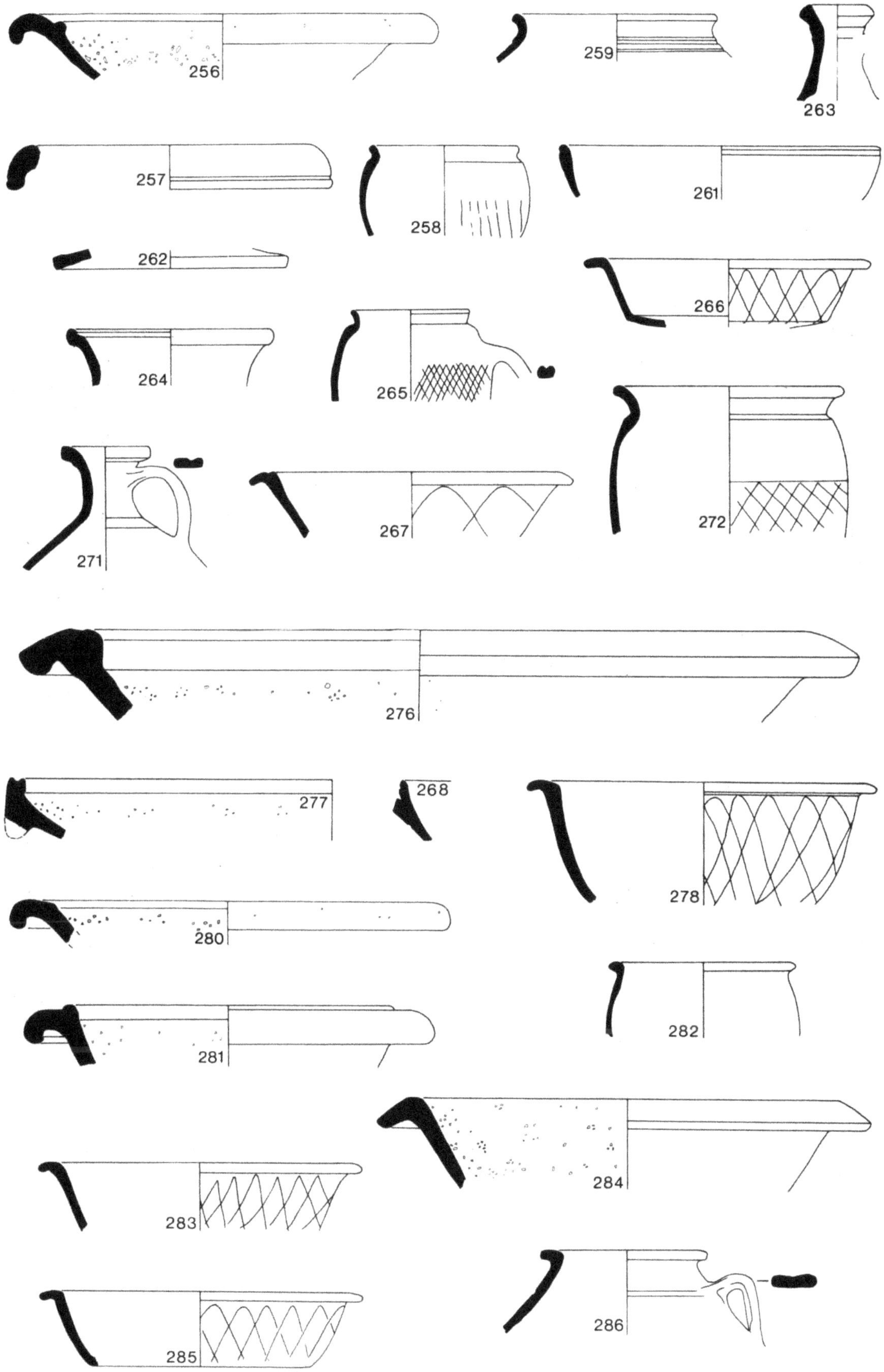

Figure 43: Pottery, scale 1:4

potteries in the first half of the second century. (K F H).

314. (Not illustrated). Flange of a mortarium in off-white fabric. Made in the Mancetter-Hartshill potteries *c* AD 160-240. (K F H).
315. Hammerhead mortarium in white fabric. A weathered fragment from a vessel made in the Mancetter-Hartshill potteries *c* AD 160-240. (K F H).
316. Flanged bowl with reeded rim in mid-grey fabric. One of the Flavian-Trajanic series. 317. Flanged bowl in orange-brown fabric.
318. (Not illustrated). Portion of a wide-mouthed jar in off-white fabric with traces of red colour coat inside and out. The vessel is broken at the shoulder but the rim appears originally to have been slightly everted. An inwardly projecting portion has been broken from the interior of the shoulder. cf Holt (Grimes 1930) no 216; Manchester (Jones and Grealey 1974) no 261. The vessel was probably a wine strainer and may be a Holt product.
319. Bowl in light orange Cheshire Plain fabric; cf Holt (Grimes 1930) no 95. Late first to early to mid second century.
320. Bowl in light orange-red Cheshire Plain fabric. An oxidized version of bowls in reduced fabric illustrated elsewhere in this report (eg no 57). (0/80).
321. Carinated bowl in orange-buff Cheshire Plain fabric. A form related to that of the flanged and carinated bowls of the Flavian-Trajanic period seems most likely. (10/43).
322. Dish in Black-burnished ware; cf Gillam 1970, no 329. AD 190-340. A hole has been pierced through the side wall, perhaps to secure a lead rivet but possibly to convert the dish for use as a cheese press.
323. Lower portion of an amphora in fabric type H. See p 96 above.

The rims of four amphorae of Dressel form 20 are illustrated. No 324 is from Stage VIII or later while the remainder are from post-complex levels. For the Martin-Kilcher types quoted below see Peacock and Williams 1986, 137-8.

324. Martin-Kilcher type 13. Mid first to early second century.
325. Martin-Kilcher type 17. Late first to early second century.
326. Martin-Kilcher type 21. Late first to early second century.
327. Martin-Kilcher type 23. Late first to early second century.
328. The vessel is in a similar fabric to that of the Pélichet 47 amphorae. Its size suggests that it may well have served as a stopper for this type of amphora. The form of this piece does, however, closely resemble that of vessels used to lighten vaulting (cf Lugli 1957, i; 671, fig 138) but it is probably too small to fulfil this task successfully; service as a stopper seems more likely.
329. A vessel in light red fabric with white slip externally. Probably part of a chimney pot, cf Lowther 1976, fig 3, no 4.

Medieval Robbing

The majority of the pottery from this phase adds little to the overall pattern of ceramics at Pentre Farm. The following pieces are, however, of interest:

330. Jar in Black-burnished ware; cf Gillam 1970, no 140. AD 160-230. (2/9).
331. Jar in smooth light red fabric with orange-buff surface; possibly Severn Valley ware and certainly related to Severn Valley ware jars such as Webster 1976, no 11. Third century. (0/25).
332. Wide-mouthed jar in Severn Valley ware (light red fabric with orange-buff surface); cf Webster 1976, no 22, 29. Second to third century. (0/61).
333. Bowl or dish in Black-burnished ware; cf Gillam 1976, nos 70-1, 73. Mid second to early third century. (2/13).
334. Mortar-like bowl in smooth orange-buff fabric. (0/22).

POTTERY FROM EARLY EXCAVATIONS AT PENTRE FFWRNDAN

By P V Webster (1976-7)

The finds from the various earlier excavations at Pentre Ffwrndan illustrate the dangers of interpreting the chronology of a whole settlement from excavations on only one or two buildings in it. The published samian evidence from each of the three campaigns (Atkinson & Taylor 1924; Petch & Taylor 1925; Petch 1936) shows a different pattern. Treatment of this material in detail would require re-examination and re-illustration but in most cases it is possible to sort the material by form and area of origin, although occasionally it is not possible to be certain of the latter (eg forms 27 and 33). Despite these doubtful cases Table 5 below shows the general outline.

The 1923 excavations produced mainly late first century samian. By contrast the 1924 excavations in the same general area produced mainly second century material, a large part of which should belong to the later half of the century. The 1932-4 material is divided between a later first century collection and a later second century one; it also includes a single example of form 29. Grouping together of all the early material does not produce a pattern identical to that presented by the 1976/77 excavations (form 18/31 is absent, for instance, and there is a dearth of earlier Central Gaulish vessels). In this way, however, a more general spread is achieved which (like that from the

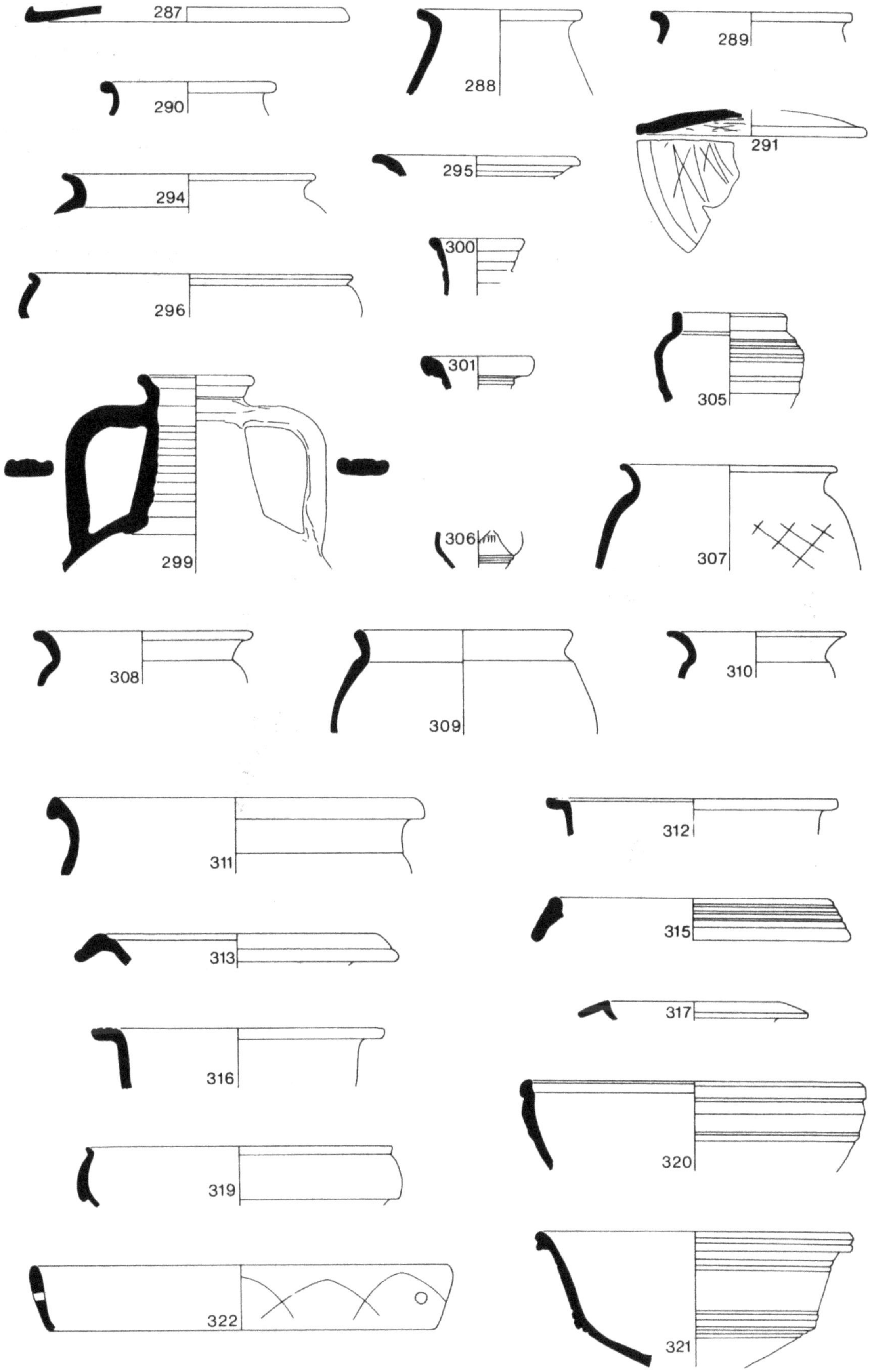

Figure 44: Pottery, scale 1:4

1976/77 excavations) goes from the late first to the late second century. As with the 1976/77 material, form 29 is scarce (only one example from 1976/77 and one from 1932-4). Clearly none of the areas of Pentre Ffwrndan examined so far has produced evidence of occupation before *c* AD 85/90.

Among the small quantity of coarse pottery published from earlier excavations it is not hard to find parallels with 1976/77 material (cf for instance, Atkinson & Taylor 1924, fig 3, no 7, or Petch & Taylor 1925, figs 11, nos 8, and 10). It seems likely that all sites were dependent upon similar potteries, as would be expected. The latest coarse pottery published appears to be the Hadrianic-Antonine jars (Atkinson & Taylor 1924, figs 11, nos 6, and 9).

Despite the lack, from earlier excavations, of any published pottery of a later date than the late second century, it is clear that there was some post-second century occupation. Coins mentioned as coming from the site (Davies 1949, Atkinson & Taylor 1924) include nine of the period *c* AD 69-212 but also five of the period *c* AD 265-353. If this coin evidence is compared with the pottery from the 1976-7 excavations, where most of the material was of the period *c* AD 90-240 but with a few late third to fourth century sherds, it would seem likely that excavations have yet to reveal the location of later occupation at the Pentre Farm settlement.

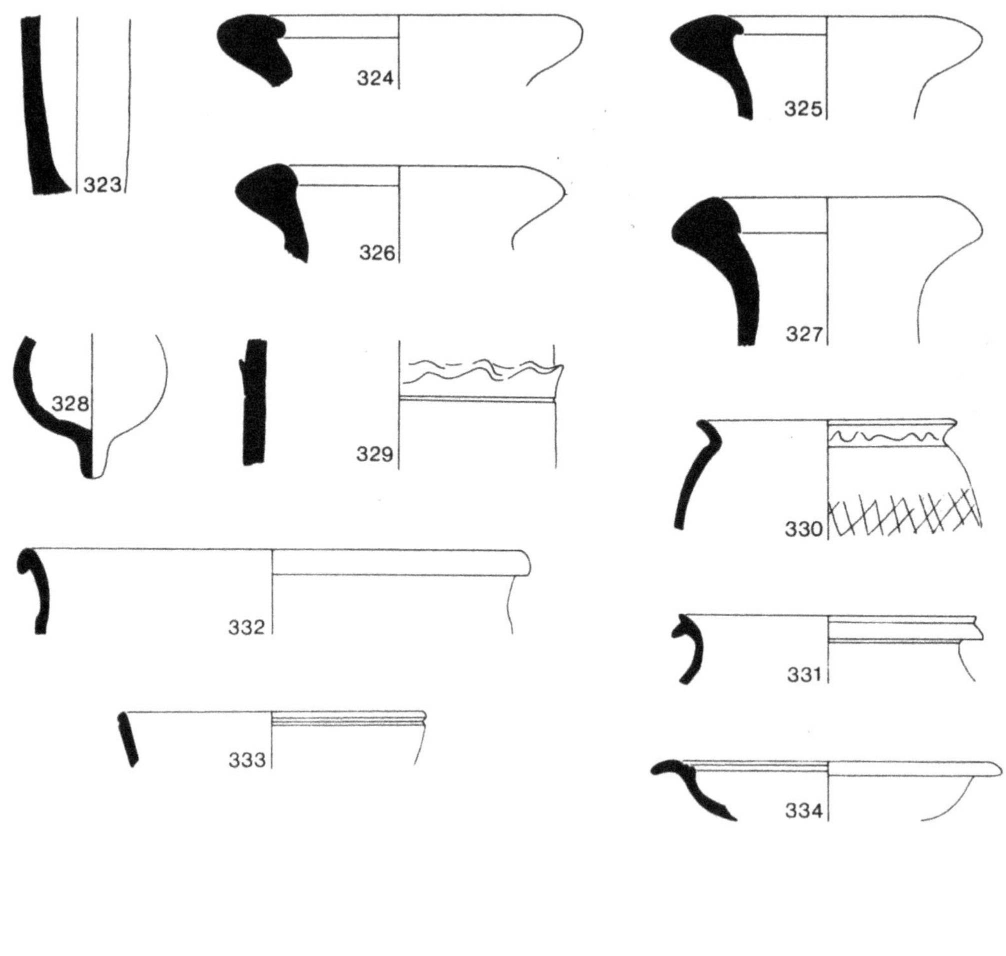

Figure 45: Pottery, scale 1:4

Table 5: Pottery from early excavations at Pentre Ffwrndan

Samian fabric		Excavations 1923	1924	1932-4	Total
Southern Gaulish					
Form	15/17	1	-	-	1
	18	3	1	6	10
	18/31	2	1?	-	3
	27	1	2?	6?	9
	29	-	-	1	1
	37	8+1?	3	12	24
Central Gaulish					
Form	30	-	-	1	1
	31	-	15	40	55
	33	1+2?	4?	6?	13
	37	-	12+1?	6	19
	38	-	1	-	1

Part IV: Specialist reports

THE ANIMAL BONES

By A King and the late B Westley (1976-7)

Of the boxes of animal bones examined about a thousand fragments were identified, representing 85% of the total. In this report the unidentified bone has not been taken into account, and about 200 rib fragments have also been excluded because of the difficulty in allotting them to their species in a mixed group. The analysis is therefore based on a sample of only 749 identified bones and fragments, about 56% of which came from insecurely stratified layers and much of which was poorly preserved, probably due to acidic soil conditions or to poor recovery. It was not possible to sieve the deposits, so there may have been a loss of bones from the smaller species such as birds and rodents (none of the latter occur in the sample). Finally, the pottery from Pentre Farm indicated a high level of residuality, and similar difficulties may be assumed of the bone, making analysis by stage virtually meaningless for such a small sample. For these reasons it has been thought best to treat the bone fragments as a single group, representing and debris that accumulated during the whole of the period between AD 90/120 and *c* AD 240. Bone from the lower levels of the topsoil has been included because no identifiable domestic activity took place after *c* AD 240. Animals present and numbers are as follows:

Mammalia Horse, ox, sheep/goat, pig, red deer, roe deer, hare, fox and cat. Goat bones were not distinguished from those of sheep; thus the two species have been grouped together.
Aves Goose, teal, chicken, crane, oystercatcher, lapwing, grey plover, woodcock and carrion crow.
Pisces Mackerel, plaice and a fish of the cod type.
Amphibia Frog.

Stages		Cow	Sheep/ goat	Pig	Horse	Red deer	Hare
I-III	BN	18	14	1	-	-	1
(early-mid/late	%	55	42	3			
second century)	MN	5	6	1			
IV-VII	BN	59	45	19	6	-	3
(late second-early	%	48	37	15			
third century)	MN	21	16	10			
VIII	BN	25	9	4	-	-	2
(early-mid third	%	66	24	10			
century)	MN	3	3	2			
Unstratified	BN	165	101	72	34	1	7
(early second-mid	%	49	30	21			
third century)	MN	25	33	21			

BN: number of fragments; MN: minimum number of individuals; percentages and MN are for the three main species only except for the total where both are given for all species

Table 6: Number of bone fragments and individuals

The number of fragments, estimated number of individuals, etc for the *mammalia* are shown in Table 6 compiled by A King. It is evident that beef provided the bulk of meat consumed, even though the number of pig and sheep carcasses passing though the kitchens may have been nearly as high.

Methods of Butchery

All parts of the skeleton seemed to be represented, except the cranium and extremities, the latter absence being most likely due to local soil conditions. There were only the smallest fragments from skulls, if whole jaws are excluded; jaws occured quite commonly, especially among the pigs (their mandibles represented a large portion of the pig bones), since the mandible is of denser bone, survives well, is conspicuous, and hence easily recovered (Grant 1971).

There was no instance of vertebral sawing or cutting to halve the carcass, but very many of the bones bear saw or chopper marks, and many were also gnawed by dogs or rodents. The lack of dog bones in the assemblage is surprising in light of this. The uniform chopping of bones, many split longitudinally, may mean that they were used for making broth, glue or grease, or to extract bone marrow from long bones. A certain amount of breakage can also occur after discarding, especially if the rubbish is redeposited later or disturbed by foraging animals. The chopping had been done methodically, with hardly any bones surviving whole among the ox, sheep and pigs, except such bones as the tarsals and carpals.

Age, disease and injury

Little indication of age range can be gained from the Pentre Farm collection. There were instances of deciduous (milk) teeth and of bones with unfused epiphyses, but with such material it was idle to speculate upon relative numbers of juveniles and adults. To the eye, it seemed that nearly all of the animals were mature adults, but not old, probably in their third year. The juveniles were well developed, past their first year but not yet fully matured. There was no evidence for the slaughter of very young animals, and except for two or three lamb bones, no first year bones at all, again possibly due to soil conditions. There was no trace of very aged animals except for one or two remains of large pig, probably boar. The flocks and herds appeared healthy, and the only indication of disease was in a horse phalanx that appeared distorted and eroded.

Discussion of Species

Ox

Whole skeletons of Romano-British oxen seem to be unknown, but published bone collections from Romano-British sites suggest that the oxen were very small animals, and the bones from the present excavation confirmed this. Later Roman remains indicated the presence of the typical Celtic ox, a small animal. Confirmation from skull evidence (long frontals and medial prominence between the horns) was lacking, however, since there were no skull parts present except for small fragments of horn core and mandibles. Three complete long bones were measured: metacarpal 182 mm overall length; metatarsal 198 mm overall length; metatarsal 168 mm overall length. These measurements resemble those of bones from other Romano-British sites (Jewell 1962). They suggest an animal not much bigger than a donkey.

Sheep/goat

Again there was no skull evidence except a few fragments of horn core, all of sheep, providing some evidence of a reasonable range of size, though none of the animals was large. The horned sheep suggested a type similar to the Soay sheep of the Western Highlands, as does the measurement of a metatarsal, the only complete main limb bone, which is 112 mm overall length. An astragalus measures only 12 mm by 9 mm (overall length and width), again suggesting a small and delicate-boned beast.

Pig

Bones fuse at irregular ages, and the Pentre Farm material gave us little information other than that it was a small domestic pig type. Again there was no skull evidence, but there were many mandibles, and here and there pieces of large tusks from boar. Wild boar, although rather difficult to distinguish, might well suggest hunting, an occupation which may have had definite economic value. In addition, wild boar may well have fertilised domesticated sows.

Deer

Red deer and roe deer are represented by six fragments, and were clearly not a mainstay of the diet. They were possibly the quarry of huntsmen, although since fragments are of antler only, perhaps the antlers were collected after being shed naturally. The red deer fragments were from a large animal. The brow tine is 300 mm long.

Horse

Fragments represent 5% of the total, slightly higher than average for Romano-British sites. Oxen would have been the usual draught animals. The horse bones bore no indication of cut or saw marks, unlike the sheep and cattle, but butchery methods may have been different for a horse. It may be that horse was eaten, if only by other animals, as is suggested by the dismembered skeletons. As often as not, the fragments consisted only of the odd tooth, but one complete astragalus indicated a quite small animal, the size of a pony. There are two complete limb bones: tibia 300 mm overall length; metatarsal 231 mm overall length. This animal may have been a mule. These measurements accord well with similar ones recorded from Verulamium and Milton Keynes, and the Romano-

British horse appears to have been small, resembling a pony of about 13 hands.

Hare

Bones occured, usually as a single fragment, in most parts of the excavation, the proportion seemingly high when compared with its occurance at other sites. It must have been a useful addition to the diet. The Romans relished hare meat and Varro mentions their domestication in hare colonies (leporaria) where they were kept for use in the kitchen (Zeuner 1963). Whether or not this was attempted in Britain is impossible to tell. The hunting of hares in the open field was a popular Roman sport.

Cat

Cat was represented by part of a jaw and a single canine tooth which seemed quite large and might have belonged to a wild cat, though it has always been an animal of woods and mountains rather than the plains. The domestic cat was rare and valuable acquisition in Roman Britain. It bred with *Felix sylvestris,* and there may have been mixed breeds and varieties quite early; little is known of the history of cats in Britain.

Fox

Bones occurred only twice.

Frogs

Bones were not numerous or remarkable.

Fish

The identification of a few small and very fragmented fish bones was the work of Mr A C Wheeler, of the British Museum of Natural History, to whom we are indebted. They revealed mackerel, plaice and a fish of the cod type (a *Gadid),* and would suggest local sea fishing.

Birds

The following list of the bird fauna is the work of Mr G Cowles of the British Museum of Natural History:

Greylag or domestic goose	*Anser anser*
Teal	*Anas crecca*
Domestic chicken	*Gallus gallus* (domestic)
Crane	*Grus grus*
Oystercatcher	*Haematopus ostralegus*
Lapwing	*Vanellus vanellus*
Grey plover	*Pluvailis squatorola*
Woodcock	*Scolopax rusticola*
Carrion crow	*Corvus corone*

Mr Cowles added the following comment: 'All the birds, with the exception of the chicken and woodcock are birds of the shoreline, mud flats, marshes and estuaries and are to be expected on a river estuary. The woodcock usually frequents woodland, but could be found close to the sea where the trees are close to the shore. The crane ceased breeding in England (East Anglia) about AD 1600 but turns up in archaeological sites (for example Valkenburg, Davies 1971, 130, table 11) and so presumably was fairly common in Roman times. The lapwing is mainly a farm bird and likes arable land, but very large flocks are often to be found in hard weather on mud flats and estuaries. Carrion crow is often to be seen feeding on the sea shore and edges of estuaries'.

The crane appears to be the first recorded on an archaeological site in Wales although it has appeared at Newstead (Davies 1971, 130). Domestic fowl remains were numerous. There were no heads or necks; the pieces were mostly of limb bones and number about eighty, with several cockspurs. In one instance, two spurs together with a tibia that had been broken and had subsequently healed are probably evidence of cockfighting.

Additional note (A K)

Bone assemblages from Roman sites in North Wales are few and far between, so the information from Pentre Farm, despite a sample size which precludes full analysis, is useful in providing much-needed data on the meat diet of the area in the early Roman period. It can be seen from Table 6 that ox and sheep/goat provide the bulk of meat supply throughout the occupation of the Pentre Farm complex, with pig bones increasing in the later stages. This is to be expected in general terms, although the low percentage of pig is unusual, which may be due either to a lack of woodland in the vicinity, or to soil-induced differential preservation on the site. If the former explanation accounts for low pig numbers, then their increase in later stages may indicate an increase in woodland in the area towards the end of the occupation of the site. The lack of wild animals and woodland bird species would appear to support the suggestion that little woodland existed, or was not exploited, at the time of occupation. However, until further information is available from the region, it cannot be said with any certainty that Pentre Farm is representative, and it may be that the use of the site has conditioned, to a greater or lesser extent, the way in which the bones reflect available animal resources.

The occupants of 'rich' sites like Pentre Farm, from which political or economic control might have been exerted over the surrounding region, would have been in a position to choose the diet that suited them, rather than having to subsist on what the local agricultural methods could produce as would have most of the poorer sites such as re-occupied hillforts (ie Dinorben) and native settlements in the area. In fact, a hypothesis could be suggested from sites like Pentre Farm, that the higher the status of a site from which a given assemblage comes, the less likely it is that the assemblage will be representative of the agricultural base of its area, due to an increasing ability to make selections from outside the local food supply.

ANALYSIS OF SEDIMENTS

By M L Shackley (1976-7)

During the main part of the excavation all Roman layers were sampled, apart from robber trench fills, using methods agreed upon with the analyst (Shackley 1975, 23-31), who saw many of the samples *in situ*. In the hurried later stages of the excavation Samples 1-4 were the only ones taken. Specific questions were posed for each sample. Dr Shackley's answers have been considered in the main part of the text and the detailed descriptions of the sediments have been retained in the archive. (T J O'L).

Method of Study

The examination of these sediments required a qualitative rather than a quantative approach, consisting of the following standard procedures:

(a) Visual examination using a hand lens.

(b) Dispersing a sub-sample in distilled water. Disaggregation using ultrasonics. Removal of the silt and clay fraction by wet sieving. Drying the coarse material under infra red light and examining it under the binocular microscope at magnifications of x6 to x50.

(c) Noting the colour, using the Munsell system.

(d) Testing sub-samples (qualitatively) for alkali-soluble humus content and presence of phosphates. Measuring pH.

General conclusions

The phosphate and humus content of all (15) samples (with the exception of Samples 1-4 and 13) were uniformly high and showed little variety. Interesting features of these sediments were:

(i) The almost universal presence of metal slag and fused fragments of glass slag (except in Samples 1, 2 and 4).

(ii) The almost universal presence of building debris (except in Samples 1-4).

(iii) The virtual absence of burning in the sediments themselves (one might have expected fractured quartz grains), despite the inclusions of much burnt material (less in Samples 1, 2 and 4).

(iv) The absence of comminuted bone fragments from many of the samples: even in samples which the excavator suspected were domestic rubbish there was very little bone. This would suggest industrial rather than domestic activity or that soil conditions were not favourable to the preservation of animal bones.

The Samples

1. 4/17, pre-complex ditch fill. A clean silty loam, free from inclusions or evidence of human activity. Approximately 10% clay: 30% silt: 60% sand. Colour 7.5YR 3/2 dark brown. Coarse fraction very small, little charcoal and no occupation debris. Fine sand grains so clean and fresh that a primary source of blown beach or dune sand would not be out of the question. The finest material (clay fraction) small, but sufficient to impart a certain cohesiveness to what would otherwise have been a loose, crumbly, open-textured, porous loam.
2. 11/61, Stage I pit fill. Loose sand mixed with less than 10% of silt and clay. Colour 7.5YR 4/4 brown, but mottled to yellowish brown. Some larger lumps of decayed organic matter. A very sandy texture with a few agglomerates caused by the small clay fraction (added from topsoil at a later date?). Little humus, much less than Sample 1, but higher phosphate content, occasional erratic and ironstone pebbles and some green glass slag together with one fragment that might have been mortar. Little charcoal.
3. 11/66, Stage II debris from possible furnace. The base consisted of a burnt loose sandy loam with fractured quartz sand grains and fire-blackened gravels. Substantial quantities of burnt bone, charcoal and slag; a test for metallic iron carried out on the lumps of slag proved strongly positive. The quartz grains had themselves been burnt, which suggested a high-temperature fire.
4. 11/54, Stage IV pit fill. Rather similar to Sample 1 but with 10% more clay. Clean textured rather sandy clay loam. Colour 7.5YR 4/4 brown. Contained brick fragments and some quite large lumps of charcoal but no metal slag. A few fragments seemed to be tile and locally-derived gravelly material.
5. 6/18, Stage IV gully fill. 7.5YR 2/0 black open-textured clay loam. Similar to Sample 7.
6. 7/14, Stage VII gully fill. 10YR 3/2 very dark greyish brown gravelly clay. Similar in many ways to Sample 9. Unburnt with quite a lot of bone fragments, tiny molten drops of metal, glass slag, charcoal and plaster.
7. 3/8, Stage VIII occupation debris. 10YR 3/2 very dark greyish brown gravelly clay. Mottled and agglomerated into thick clay balls. The clay had been contaminated with debris but the natural coarse fraction (sand and gravel) was very small. Fragments of slate, brick, mortar and glass slag.
8. 0/1, post-complex destruction level. Very little sediment, mostly pottery and burnt building debris. Large fragments of charcoal and burnt plaster.
9. 4/1, debris from demolition between Stages VII and VIII, also the western ditches. 10YR 2/1 black, loose crumbly loam with much brick, not rich in clay. Not well agglomerated. Glass slag, burnt mortar, clay, plaster and bone.
10. 4/2, Stage VII possible debris from bath-house flues lying in ditch to the west of the complex. 7.5YR 2/0 black. Similar to Sample 9.
11. 12/8, Stage VII pit fill; from the stratification to

the north of the complex. Richer in clay. Contaminated topsoil with many modern roots. 10YR 1/2 black. Much brick in the sample. Charcoal, some fine sand, locally derived pebbles and some mortar. Sediment itself unburnt but burnt debris included. Minute fragments of metal slag and glass but no bone. Quartz grains themselves show no sign of having been burnt.

12. 12/13, Stage VII pit fill; from the stratification to the north of the complex. Very similar in colour (10YR 2/2 very dark brown) to Sample 13 but with a looser more crumbly texture since it contains less clay. Less root debris. Silt and clay content less than a third of that of Sample 13, but the humus and phosphate reactions stronger (redeposited topsoil ?). One lump of mortar, but otherwise very little building debris. Green glassy fragments. Quartz sand grains very well rounded and frosted, many times redeposited.
13. 12/24, Stage V tank fill; from the stratification to the north of the complex. 10YR 2/2 very dark brown clay-rich sediment. Agglomerated and with less sand and clay and more silt than Sample 14. Includes much brick and large lumps of charcoal. Humus reading very strong, phosphates not so strong as Sample 14. Large lumps of charcoal from large pieces of wood, whereas the charcoal from Sample 14 is from smaller twigs. Glass slag, mortar and brickwork.
14. 12/17, Stage V tank fill; from stratification to the north of the complex. 7.5YR 4/4 dark brown charcoal, modern roots and angular flint pebbles. Mottled appearance with patches of lighter material. Similar to sample 13 but without the fine (silt and clay) fraction the sandy decayed brick, burnt material and charcoal. Fragments of glass slag, metallurgical slag but no bone or shell. Quartz grains very rounded and frosted but clean and unburnt. Redeposited many times. High humus and phosphate content.
15. 12/15, Stage uncertain. Destruction debris to the north of the complex. 7.5YR 2/0 black. Rather sandy in texture with vast quantities of burnt brickwork, glass fragments, metal slag, burnt mortar and plaster.

FLORAL AND FAUNAL REMAINS

By Kenneth Brassil (1976-7)

The Mollusca

The molluscan remains have been identified as belonging to thespecies shown in Table 7.

These data do not necessarily reflect the role of shellfish in the economy of the site during its occupation. The vicissitudes of preservation and excavation (most of the bivalves were broken and hence difficult to collect) will necessarily cause the recorded data to deviate from the actual numbers of molluscs, though it is evident that shellfish were a component of the diet during various periods of settlement.

The species of marine molluscs recorded do occur in estuarine habitats; in modern times, however, the distribution of *O edulis* in particular has suffered from the ravages of pollution and predation, to the extent that oyster beds are not common around the coasts of Britain. During the Roman period they may have been much more widespread. The oysters may also have been obtained from localities other than that of the Dee estuary, or may even have been imported (Thomas 1974).

It is interesting to note the presence, albeit minor, of *H aspersa* (all large specimens). This is an edible snail whose culinary attributes have been praised by Pliny. Evans (1972) records the presence of this species in Roman contexts.

Species	Total abundance	% occurrence
Mytilus edulis mussel	17	6
Cardium (Cerastoderma) edule cockle	12	4
Ostrea edulis oyster	191	70
Helix aspersa Muller garden snail	14	4

Table 7: Molluscan remains

Analysis for Seed Remains

Thirty soil samples varying between 0.5 and 1.0 kg were treated with dispersant (Calgon) and either air or oven dried at low temperature. The samples were then sieved and viewed under a binocular microscope and floral and faunal remains were identified and collected.

The majority of the samples were composed of soils having a high charcoal content. It was hoped that such deposits would preserve burnt and charred seeds. Results from small samples (in individual volume and total numbers) indicate that seed remains were often either not present initially or not preserved subsequently. One sample contained a single specimen of *Agrostemma githago* (Linnaeus) - the corn cockle - a weed usually associated with cereals. Examples have been discovered at *Isca* (Helbaek 1964). Two other samples contained six fragmentary specimens of burnt seeds. Identification to generic level was not possible.

It is suggested that the aerobic conditions prevalent in the deposits precluded the preservation of floral remains in quantity.

Bibliography

Abbreviations

The abbreviations used follow the recommendation made in the CBA's standard list of abbreviated titles, with the addition of the following:

AA (4)	Archaeologia Aeliana, ser 4
BAR	British Archaeological Reports
CBA	Council for British Archaeology
D.	Déchelette, J, 1904
O.	Oswald, F, 1936-7
Rogers	Rogers, G B, 1974
S & S	Stanfield, J A, & Simpson, G, 1958
WAT	Western Archaeological Trust

Allen, D, 1986 The glass vessels, in Zienkiewicz, J D, 1986, 96-116.

Allen, J C, 1970 Considerations on the antiquity of mining in the Ibernian Penninsula, *R Anthropol Inst Occ Papers*, **27**.

Anderson, A C, 1981 Some continental beakers of the first and second centuries AD, in Anderson, A C, & Anderson, A S, 1981, 321-47.

Anderson, A C, & Anderson, A S, 1981 *Roman pottery research in Britain and beyond*, BAR, **S123**.

Atkinson, D, 1942 *Report on Excavations at Wroxeter (the Roman City of Viroconium) in the County of Salop 1923-7*, Birmingham Archaeol Soc.

Atkinson, D & Taylor, M V, 1924 Flint excavations report, *J Flintshire Hist Soc*, **10**, pt 1, 5-23.

Arthur, P, & Marsh, G (eds), 1978 *Early fine wares in Roman Britain*, BAR, **57**.

Avent, R, & Howlett, T, 1980 Excavations in Roman Long Melford 1970-1972, *Proc Suffolk Inst Archaeol*, 34, pt iv, 229-49.

Baillie Reynolds, P K, 1938 *Excavations on the site of the Roman Fort of Kanovium, at Caerhun, Caernarvonshire*. Cardiff.

Barbet, A, 1973 Remontage des peintures murales romaines, Recherches d'Archéologie Celtique et Gallo-Romaine, in Duval, P-M, (ed) 1973, 67-81.

Barbet, A, 1974 Recueil général des peintures murales de la Gaule, i, Province de Narbonnaise, 1, *Gallia Supplement*, **27**.

Bender, H, 1975 Archäologische Untersuchungen zur Ausgrabung Augst-Kurzenbettli, ein Beitrag zur Erförschung der römischen Rasthäuses, *Antiqua*, **4**. Bender, H, 1978 Römische Reiserverkehr, Stuttgart.

Berger, L, 1960 *Römische Gläser aus Vindonissa*, Basel.

Besuver, M, 1921 Le commerce du plomb á l'époque Romaine, d'aprés les lingots estampilles, *Revue Arch*, **13**.

Biddle, M, 1967 Two Flavian burials from Grange Road, Winchester, *Antiq J*, **47**, 224-50.

Bird, J, et al (eds) *Southwark excavations 1972-4, II*.

Bird, J, et al (eds) 1978 *Collectanea Londiniensia:* Studies presented to Ralph Merryfield, London Middlesex Archaeol Soc, Spec Pap, 2.

Birley, E, 1935 Fourth report on excavations at Housesteads, *AA (4)*, **12**.

Birley, E, 1953 *Roman Britain and the Roman Army*.

Birley, E, Dobson, B, & Jarrett, M, 1974 *Roman Frontier Studies*.

Bishop, M E, (ed) 1985 *The Production and Distribution of Roman Military Equipment*, BAR, **S275**.

Blake, M E, 1947 *Ancient Roman Construction in Italy from the Prehistoric Period to Augustus*, Carnegie Inst, Washington.

Blake, M E, 1973 *Roman Construction in Italy from Nerva through the Antonines*, edited and completed by D T Bishop, Mem Amer Phil Soc, **96**, Philidelphia.

Blay, W, 1925 *Letocetum*.

Blockley, K, forthcoming *Prestatyn; Melyd Avenue Excavations 1984-5*, BAR.

den Boesterd, M H P, 1956 *The bronze vessels in the Rijksmuseum G M Kam at Nijmegen*, Nijmegen.

Bogaers, J E, 1964 Praetorium Agrippinae, *Bull Koninklyke Nederlandsche Oudheid-Kundige Bond*, **6** ser, 17.

Boon, G C, 1966 Roman window glass from Wales, *Journal of Glass Studies*, **8**, 41-5.

Boon, G C, 1967 Roman glass from Wales, *Annales du 4me Congrès des "Journées Internationales du Verre"*, 93 102.

Boon , G C, 1972-3 Roman Glassware from Caerwent

1855-1925, *Monmouthshire Antiq*, **III**, part ii, 111-123.

Boon, G C, 1974 *Silchester: The Roman Town of Calleva.*

Boon, G C, & Savory, H N, 1975 A Silver Trumpet-Brooch with Relief Decoration, Parcel-Gilt, From Carmarthen, and a note on the Development of the Type, *Antiq J*, **55**, pt 1, 41-61.

Brailsford, J, 1962 *Hod Hill 1. Antiquities from Hod Hill in the Durden Collection*, British Museum.

Branigan, K, & Fowler, P, 1975 *The Roman West County.*

Brewer, R J, 1986 The beads and glass counters, in Zienkiewicz, J D, 1986, 146-56.

Brewster, N H, 1972 Corbridge: its significance for the study of Rhenish ware, *AA (4)*, **50**, 205-16

Brill, R, & Wimpler, J, 1967 Isotope studies of ancient lead, *American J Archaeol*, **71**.

British Museum, 1920 *A Guide to the Exhibition illustrating Greek and Roman life*, 2nd edn.

British Museum, 1922 *A Guide to the Antiquities of Roman Britain.*

British Museum, 1964 *Guide to the Antiquities of Roman Britain.*

Brodribb, ACC, Hands AR, & Walker DR, 1971 *Excavations at Shakenoak, II*, privately printed.

Brodribb, G, 1987 *Roman Brick and Tile.*

Brothwell, D R, 1967 *Digging up bones.*

Brunt, P, 1966 *Procuratorial Jurisdiction*, Brussels.

Bushe-Fox, J P, 1913 *Excavations on the site of the Roman town at Wroxeter, Shropshire, in 1912*, Res Rept Soc Antiq of London, **I**.

Bushe-Fox, J P, 1949 *Fourth report on the excavations of the Roman fort at Richborough, Kent*, Res Rept Soc Antiq of London, **XVI**.

Butcher, S A, 1967 The Finger Rings in Dudley, D, 1967, 21f.

Butcher, S A, 1974 The brooches, in Neal, D S, 1974, 123-8.

Carrington, P, 1977a The planning and date of the Roman Legionary fortress at Chester, *J Chester Archaeol Soc*, **60**, 35-42.

Carrington, P, 1977b Severn Valley ware and its place in the Roman pottery supply at Chester: a preliminary assessment, in Dore, J, & Greene, K, (eds) 1977, 147-62.

Cerskov, E, 1970 *Municipium D D kod Socvanica*, Belgrade.

Chaplin, J, forthcoming *Excavations in Dorchester*, Dorset Nat Hist and Archaeol Soc Monograph.

Charlesworth, D, 1959 Roman glass in northern Britain, *AA (4)*, **37**, 33-58.

Charlesworth, D, 1966 Roman square bottles, *Journal of Glass Studies* , **8**, 26-40.

Charlesworth, D, 1971 A group of vessels from the Commandant's house, Housesteads, *Journal of Glass Studies*, **13**, 34-7.

Charlesworth, D, 1975 The Commandant's House, Housesteads, *AA (5)*, **3**, 17-42.

Charlesworth, D, 1980 Glass. in Stead, I M, 1980, 124-5.

Charlesworth, D, 1981 Glass from the burials, in Partridge, C, 1981, 268-71.

Chevalier, R, 1976 *Roman Roads.*

Cleere, H, 1978 *Roman Harbours in Britain South of Hadrian's Wall.*

Collingwood, R, & Richmond, I A, 1969 *The Archaeology of Roman Britain.*

Colls, D, et al, 1977 'L'épave Port-Vendres II et le commerce de la Bétique á l époque de Claude, *Archaeonautica*, **I**, Paris.

Comstock, M B, & Vermeule, C, 1971 *Greek Etruscan and Roman Bronzes in the Museum of Fine Arts*, Boston.

Cool H, E, M, & Price, J, forthcoming The Roman Glass, in Chaplin, J, forthcoming.

Corder, P, 1935 *Excavation at the Roman fort at Brough, E Yorkshire, 1934*. Hull Univ Coll, Local Hist Comm.

Corder, P, 1961 *The Roman Town and Villa at Great Casterton, Rutland. 3rd Rept for the years 1954-8.*

Cunliffe, B W, 1968 *Fifth report on the excavation of the Roman fort at Richborough, Kent*, Res Rept Soc Antiq of London, **XXIII**.

Cunliffe, B W, 1971 *Excavations at Fishbourne, 1961-9, II, The Finds*, Res Rept Soc Antiq of London, **XXVII**.

Cunliffe, B W, 1974 *Iron Age Communities in Britain.*

Cunliffe, B W, 1975 *Excavations at Portchester Castle, Vol I, Roman.* Res Rept Soc Antiq of London, **XXXII**.

Curle, J, 1911 *A Roman Frontier Post and its People, The Fort of Newstead in the Parish of Melrose.*

Daniels, C, 1959 Roman Bath-house at Red House, Corbridge, *AA (4)*, **37**, 85-176.

Daniels, C M, Jones, G D B, & Putnam, W G, 1971 Excavations at Caersws, 1968, *Montgomery Collect*, **61**, 37-42 (1969-70).

Daniels, C, Jones, G D B, & Putnam, W, 1970 A note on the Roman Bath-houses at Prestatyn and Tremadoc, *Bull Board Celtic Stud*, **23**, pt 2.

Davies, E, 1949 *The Prehistoric and Roman Remains of Flintshire.*

Davies, J L, 1977 Roman arrowheads from Dinorben and the Sagittarii of the Roman army, *Britannia*, **viii**, 257-70.

Davies, O, 1935 *Roman Mines in Europe.*

Davies, R W, 1974 The daily life of a Roman Soldier under the principate, *Aufsteig und Niedergang der Römische Welt*, **II**, Berlin.

Déchelette, J, 1904 *Les vases céramiques orné de la Gaule Romaine*, **2**, Paris.

Dembeck, H, 1966 *Animals and Man.*

Domergue, C, 1970 Les exploitations auriféres du nord-ouest de la peninsule Ibérique sous l'occupation Romaine, *La Minera Hispanae Iberoamericana*, **I**.

Dore, J, & Greene, K, (eds) 1977 *Roman pottery studies in Britain and beyond,* BAR, **S30**.

Droop, J P, & Newstead, R, 1928 Excavations in the Deanery Field, Chester, 1928, pt II, The Finds, *Annals of Archaeol and Anthrop,* **18**, pts 3-4, 113 ff.

Dudley, D, 1967 Excavations on Nor'nour in the Isles of Scilly, 1962-6, *Archaeol J,* **124**, 1-64

Dusavic, S, 1977 Aspects of mining in Noricum, Pannonia, Dalmatriu and Moesia Superia, *Aufsteig und Niedergang der römische Welt,* Berlin.

Duval, P-M, (ed) 1973 Recherches d'archéologie celtique et gallo-romaine, *Hautes études du monde gréco-romain,* **5**.

Eggers, R, 1966 *Das Praetorium als Amtssitz und Quartier römischer Spitzenfunktionare,* Osterreichische Akademie des Wissenschaft, Philosophie-Historie Klasse.

Elkington, H, 1975 The Mendip lead mines, in Branigan, K, & Fowler, P, 1975.

Evans, J G, 1972 *Land snails in Archaeology.*

Fellmann, R, 1958 Die Principia des Legionslagers Vindonissa und das Zentralgebäude der römischen lager und Kastelle, *Jahresber Ges pro Vindonissa,* 1957/8.

Fentress, E B W, 1979 *The Roman Army in Numidia,* BAR, **S53**.

Ffoulkes, W, 1856 Roman remains and ancient lead works, *Archaeol Cambrensis,* (3), **11**.

Fischer, U, 1973 Grabungen im römischen Steinkastell von Heddernheim, 1957-9, *Schriften des Frankfurter Museums für Vor-und Frühgeschichte,* **2**.

Forster, R H, & Knowles, W H, 1912 Corstopitum: report on the excavations in 1911, *AA (3),* **8**, 137-263.

Foster, I, & Daniel, G, (eds) 1965 *Prehistoric and Early Wales.*

Fox, A, & Ravenhill, W, 1972 The Roman fort at Nanstallon, Cornwall, *Britannia,* **iii**, 56-111.

de Franciscis, A, 1974 *Ercolano e Stabia,* Novara.

Frere, S S, 1967 *Britannia.*

Frere, S S, 1972 *Verulamium excavations, I,* Res Rept Soc Antiq London, **XXVIII**.

Frere, S S, & St. Joseph, J K, 1974 The Roman Fortress at Longthorpe, *Britannia,* **v**, 1-129.

Fulford, M, 1975 The Pottery in Cunliffe, B W, 1975, 270-367.

Gage, J, 1834 ... Barrows called the Bartlow Hills, in the parish of Ashdon, in Essex, with an account of Roman sepulchral relics recently discovered in the lesser Barrows, *Archaeologia,* **25**, 1-23.

Gibson, J P, & Simpson, F G, 1911 The milecastle on the Wall of Hadrian at the Poltross Burn, *Trans Cumberland Westmorland Antiq Archaeol Soc,* **11**, 390-461.

Gillam, J P, 1970 *Types of Roman Coarse Pottery Vessels in Northern Britain,* 3rd edn.

Gillam, J P, 1976 Coarse fumed ware in North Britain and beyond, *Glasgow Archaeol J,* **4**, 57-80.

Gowland, W, 1900 Remains of a Roman silver refinery, *Archaeologia,* **57**, pt 1.

Grant, A, 1971 The bird bones, in Cunliffe, B W, 1971, 377-88.

Green, H J M, 1976 Godmanchester, *Small Towns of Roman Britain.*

Greene, K, 1977 Legionary pottery and the significance of Holt, in Dore, J, & Greene, K, (eds) 1977, 113-32.

Greene, K, 1978a Imported fine wares in Britain to AD 250: a guide to identification, in Arthur, P, & Marsh, G, (eds) 1978, 15-30.

Greene, K, 1978b Roman trade between Britain and the Rhine Provinces: the evidence of pottery to c AD 250, in du Plat Taylor, J, & Cleere, H, (eds) 1978, 52-8.

Greene, K, 1979 *Report on the excavations at Usk, 1965-76. The pre-Flavian fine wares.*

Grenier, A, 1934 *Manuel d'archéologie gallo-romaine,* **II**, Paris.

Grimes, W F, 1930 The works depot of the Twentieth Legion at Castle Lyons, Holt, Denbighshire, *Y Cymmrodor,* **41**.

Guido, M, 1978 *The glass beads of the Prehistoric and Roman periods in Britain and Ireland,* Res Rept Soc Antiq of London, **35**

Hagen, J, 1931 *Römerstrassen der Rheinprovinz,* Bonn.

Hanson, W S, 1971 The kiln group, Northwich, in Jones, G D B, 1971, 50-3.

Harden D B, 1936 The Glass, in Radford, C A R, 1936, 26-69. The Roman villa at Ditchley, Oxfordshire. *Oxoniensia,* **I**, 26-69.

Harden D B, 1947 Roman glass. In Hawkes C F C & Hull M R, *Camulodunum.* Res Rept Soc Antiq of London, **15**, 287-306.

Harden, D B, 1962 Roman glass in York, *Eburacum - Roman York,* HMSO, 136-141.

Harden D B, 1967 The glass jug, in Biddle, M 1967, 238-40.

Harden, D B, 1971 Glass in Brodribb, A C C, et al, 1971, 98-108.

Harden D B, 1974 Window glass from the Romano-British bath-house at Garden Hill, Hartfield, Sussex, *Antiq J,* **54**, 280-1.

Harden, D, B, 1977 Report on the glass bowl, in Partridge, C, 1977, 22-108.

Harden D B, 1983 The glass hoard, in Johnson, S, 1983, 81-89.

Harden D B, & Green C M, 1978 A late Roman grave-group from the Minories, Aldgate, in Bird J, et al, (eds) 1978, 163-175.

Harden D B, & Price, J, 1971 The glass, in Cunliffe B, W, 1971, 317-68.

Harden D B, et al, 1968 *Masterpieces of glass.* London.

Harden, D B, et al 1987 *Glass of the Caesars,* Milan.

Hartley, B R, & Kaine, K F, 1954 Heronbridge excavations: Roman dock and buildings, *J Chester*

Archaeol Soc, **41**, 15-37.

Hartley, K F, & Webster, P V, 1973 Romano-British pottery kilns near Wilderspool, *Archaeol J*, **130**, 77-103.

Hartley, K F, 1977 Two major potteries producing mortaria in the first century AD, in Dore, J, & Greene, K, (eds) 1977, 5-17

Hartley, K F, 1981 Painted fine wares made in the Raetian workshops near Wilderspool, Cheshire, in Anderson, A C, & Anderson, A S, 1981, 471-9.

Hassall, M, & Rhodes, J, 1974 Excavations at the New Market Hall, Gloucester 1966-7, *Trans Bristol Gloucs Archaeol Soc*, **93**, 15-100.

Haverfield, F, 1906 *Romano-British Remains*.

Hayes, P, 1961 Flint: The discoveries near the Roman site at Pentre, *J Flintshire Hist Soc*, **19**, 93-5.

Heighway, C, 1983 *The east and North Gates of Gloucester*, WAT Monograph, **4**.

Helbeck, H, 1964 The Isca Grain, A Roman plant introduction in Britain, *New Phytologist*, **63**, 158-64.

Hermet, F, 1934 *La Graufesenque, Planches*, pt 1, Paris.

Hobley, B, 1969 A Neronian-Vespasianic military site at The Lunt, Baginton, Warwickshire, *Trans Birmingham Warwickshire Archaeol Soc*, **83**, 65-129.

Hobley, B, 1974 Excavations at The Lunt Roman military site, Baginton, Warkwickshire 1968-71, 2nd Interim report, *Trans Birmingham Warwickshire Archaeol Soc*, **85**, 7-92.

Hogg, A H A, 1966 Native settlement in Wales, in Thomas, C, (ed) 1966, 28-38.

Hogg, A H A, 1968 Pen Llystyn, a Roman fort and other remains, *Archaeol J*, **125**, 101-92.

Hull, M R, 1967 The Nor'nour Brooches, in Dudley, D, 1967, 28-64.

Hurst, H, 1972 Excavations at Gloucester 1968-71, *Antiq J*, **52**, 24-69.

Isings, C, 1957 *Roman glass from dated finds*, Groningen.

Isings, C, 1961 *Roman glass in Limburg*, Groningen.

Jacobi, L, 1897 *Das Römerkastell Saalburg*, Hamburg.

Jacobs, J, 1912 Sigillatafunde aus einen römische Keller zu Bregenz, *Jahrbuch für Altertumskunde*, **6**, 172-84 & Taf I-IV.

Jarrett, M G, 1967 The Roman Frontier in Wales, *Studien zu den Militärgrenzen Roms*, Bonn. (Beihefte der Bonner Jahrbuch **19**) 1967, 21-31.

Jarrett, M, G, 1976 Maryport, Cumbria: a Roman fort and its garrison, *Trans Cumberland & Westmorland Antiq Archaeol Soc*, Extra Ser, **22**.

Jarrett, M G, & Mann, J C, 1968 The military occupation of Roman Wales, *Bull Board Celtic Stud*, **20**, 206-20.

Johnson, S, 1983 Burgh Castle: excavations by C Green, 1958-61, *E Anglian Archaeol*, **20**.

Jones, A H M, 1974 The *Roman Economy*.

Jones, G D B, & Bird, D G, 1972 Roman gold mines in north-west Spain, pt 2, *J Roman Stud*, **62**, 59-74.

Jones, G D B, & Grealey, S, 1974 *Roman Manchester*.

Jones, G D B, & Lewis, P V, 1960 Roman gold mines in north-west Spain, pt 1, *J Roman Stud*, **60**, 159-85.

Jones, G D B, & Lewis, P V, 1971 The Dolocauthi gold mines, *Bonner Jahrbucher*, **7**.

Jones, G D B, & Webster, P V, 1968 Excavations at Whitchurch, Shropshire, 1965-6, *Archaeol J*, **125**, 193-254.

Jones, G D B, (et al) 1971 Excavations at Northwich (Condate), *Archaeol J*, **128**, 31-77.

Jope, W, 1961 *Studies in Building History*.

Kalinka, E, 1906 Antika Denkmäler in Bulgarien, *Schriften der Balkenkommission*, **4**.

Kelly, R, 1976 Metal-working in North Wales during the Roman Period, *Bull Board Celtic Stud*, **27**, pt 1.

Kleburg, T, 1957 *Hôtels, restaurants et cabarets dans l'antiquité romaine*, Uppsala.

Knorr, R, 1919 *Töpfer und Fabriken verzierter Terra-Sigillata des ersten Jahrhunderts*, Stuttgart.

Knorr, R, 1952 *Terra-Sigillata-Gefässe des ersten Jahrhunderts mit Töpfernamen*, Stuttgart.

Knott, R J, 1913 Letocetum excavations 1912, *Trans North Staffs Field Club*, **47**.

Koethe, H, 1936 Strassendorf und Kastell bei Junkerath, *Trierer Zeitschrift*, **11**.

Lambrick, G, 1980 Excavations at Park Street, Towcester, *Northamptonshire Archaeol*, **15**, 35-118.

Lawson, A, 1976 Shale and jet armlets from Silchester, *Archaeologia*, **105**, (1975), 241-75.

Liversidge, J, 1955 *Furniture in Roman Britain*.

Liversidge, J, 1968 *Britain in the Roman Empire*.

Liversidge, J, 1969 Furniture and interior decoration, in Rivet, A L F, 1969, 127-72.

Lloyd-Morgan, G, 1978 Some small Roman bronzes from the Grosvenor Museum, *J Chester Archaeol Soc*, **61**, 25-32.

Lorenz, H, 1936 *Untersuchung zum Prätorium*, Berlin.

Lowther, A W G, 1976 Romano-British chimney-pots and finials, *Antiq J*, **56**, 35-48.

Lugli, G, 1957 *La Tecnica Edilizia Romana*, Roma.

MacDonald, G, & Curle, A O, 1929 The Roman Fort at Mumrills, near Falkirk, *Proc Soc Antiq Scot*, **63**, 396-575.

MacKendrick, P, 1970 *The Romans on the Rhine*, New York.

Mackreth, D, forthcoming The brooches, in Blockley, K, forthcoming.

MacMullen, R, 1959 Roman Imperial Building in the Provinces, *Harvard Studies in Classical Philology*, Boston.

Maiuri, A, 1958 *Ercolano: I Nuovo Scavi 1927-58*, **1**, Roma.

Manning, W H, 1972 The iron objects, in Frere, S S, 1972, 163-95.

Manning, W H, 1974 Objects of iron, in Neal, D S,

1974, 157-87.

Manning, W H, 1976, *Catalogue of Romano-British Ironwork in the Museum of Antiquities*, Newcastle upon Tyne.

Manning, W H, (ed) forthcoming Excavations at Usk 1965-76.

Marshall, F H, 1907 *Catalogue of the Finger Rings in the British Museum. Greek, Roman and Etruscan.*

Meiggs, R, 1973 *Roman Ostia*, 2nd edn.

Mócsy, A, 1974 *Pannonia and Upper Moesia.*

Nash-Williams, V E, 1969 *The Roman Frontier in Wales*, 2nd edn, by Jarrett, M G.

Neal, D S, 1974 *The excavation of the Roman villa in Gadebridge Park, Hemel Hempstead, 1963-8*, Res Rept Soc Antiq of London, **XXXI**.

Neusteping, E J, 1961 *Czechoslovakia.*

Newstead, R, & Droop, J P, 1936 Excavations in the Deanery Field and Abbey Green, 1935, *Liverpool Annals Archaeol Anthropol*, **23**, 3-50.

Noeske, H-C, 1977 Studien zur Verwaltung und Bevolkerung der dakischen Goldsberqwerke, *Bonner Jahrbucher*, **177**.

Ogilvie, R, & Richmond, I, 1967 *Cornelii Taciti de Vita Agricolae.*

O'Leary, T, & Davey, P, 1977 Excavations at Flint, 1976-7 (the post-medieval period with a summary of the Roman Sequence), *J Flintshire Hist Soc*, **27**, 138-51.

Oliver, A, 1984 Early Roman faceted glass, *Journal of Glass Studies*, **26**, 35-58.

Orton, C R, 1975 It's later than you think: a statistical answer to an archaeological problem, *London Archaeol*, **2**, 285-7.

Orton, C R, 1979 Statistical dating for the sequence from trench 2, Southwark Excavations 1972-4, in Bird, J, et al, (eds) 460-1.

Oswald, F, 1936-7 *Index of Figure Types on Terra Sigillata.*

Oswald, F, & Pryce, T D, 1920 *An Introduction to the Study of Terra Sigillata.*

Partridge, C, 1977 Excavations and Fieldwork at Braughing, 1968-73, *Hertfordshire Archaeol*, **5**.

Partridge, C, 1981 *Skeleton Green*, Britannia Monograph Series, 2.

Peacock, D P S, 1977 Bricks and tiles of Classis Britannica, *Britannia*, **viii**, 235-48.

Peacock, D P S, & Williams, D F, 1986 *Amphorae and the Roman economy.*

Pennant, T, 1778 *Tours in Wales, Vol 1.*

Percivál, J, 1976 *The Roman Villa.*

Petch, D F, 1970 Excavations at the Old Market Hall, Chester: Second Summary Report, 1968-70, *J Chester Archaeol Soc*, **57**, 3-26.

Petch, J A, 1928 Excavation at Benwell, *AA (4)*, **5**.

Petch, J A, 1936 Excavations at Pentre Ffwrndan near Flint 1932-4, *Archaeol Cambrensis*, **91**, 74-92.

Petch, J A, & Taylor, M V, 1925 Report on the excavations carried out at Pentre, Flint during September 1924, *J Flintshire Hist Soc*, **10**, pt 2, 5-29.

von Petrikovits, H, 1975 *Die Innenbauten römischer Legionslager während der Prinzipätszeit*, Opladen.

Pflaum, H-G, 1975 *La fortification de la ville d'Adraha d'Arabie, Sapien*, **29**.

Phillips, B, & Walters, B, 1977 A Mansio at Lower Wanborough, Wiltshire, *Britannia*, **viii**, 223-28.

Pitts, L, F, & St Joseph, J, K, 1985 *Inchtuthil*, Britannia Monograph, **6**.

du Plat Taylor, J, & Cleere, H, (eds) 1978 *Roman shipping and trade: Britain and the Rhine Provinces*, CBA Res Rept **24**.

Price, J, 1976 Glass, in Jarrett M, G, 1976, 49-54.

Price, J, 1980 The Roman glass, in Lambrick, G, 1980, 35-118.

Price, J, 1983 The Roman vessel glass, in Heighway C, 1983, 168-70.

Price, J, 1985 Chapter 30. The Roman glass, in Pitts, L, F, & St Joseph, J, K, 1985, 303-12.

Price, J, 1987 The glass from Felmongers, Harlow, in Essex: a dated deposit of vessel glass found in an Antonine pit, *Annales du 10me Congrès de l'Association Internationale pour l'Histoire du Verre*, 185-206.

Price, J, forthcoming, The Roman glass, in Quinnell, H, forthcoming.

Quinnell, H, forthcoming Excavations at Cosgrove Roman Villa.

Radford, CAR, 1936 The Roman Villa at Ditchley, Oxfordshire, *Oxoniensia*, **I**, 26-69.

Ramsay, A M, 1925 The speed of the Roman Imperial Post, *J Roman Stud*, **15**, 60-74.

Reece, R, (ed) 1977a *Burial in the Roman World*, CBA Res Rept, **22**.

Reece, R, 1977b Burial in Latin Literature: two examples, in Reece, R, 1977a, 44-5.

Richmond, I A, 1939 Fendoch, *Proc Soc Antiq Scot*, **73**, 110-54.

Richmond, I A, 1968 *Hod Hill, Vol 2, Excavations, 1951-8.*

Richmond, I A, Romans, R T, & Wright, R P, 1944 A civilian Bath-house of the Roman period at Old Durham, *AA (4)*, **22**, 1-21.

Richmond, I A, & Gillam, J P, 1950 Excavations on the Roman site at Corbridge, *AA (4)*, **28**, 152-201.

Richmond, I A, & Gillam, J P, 1951 The temple of Mithras at Carrowburgh, *AA (4)*, **31**, 1-92.

Richmond, I A, & Gillam, J P, 1953 Buildings of the 1st and 2nd centuries north of the granaries at Corbridge, *AA (4)*, **31**, 205-53, (1953).

Richmond, I A, & Gillam, J P, 1955 Some excavations at Corbridge 1952-4, *AA (4)*, **33**, 218-52.

Richter, G M A, 1966 *Furniture of the Greeks, Etruscans and Romans.*

Ritterling, E, 1912 Das Frühromische Lager bei Hofheim in Taunus, *Annalen des Vereins für Nassauische Altertumskunde und Geschichtsforschung*, **40**,

1-416.

Rivet, A L F, 1969 *The Roman Villa in Britain.*

Robertson A, S, 1975 *Birrens (Blatobulgium)*, Edinburgh

Rogers, G B, 1974 Poteries sigilleées de la Gaule centrale, I: les motifs non figurés, *Gallia Suppl*, **28**, Paris.

Rook, A, 1976 *Aspects of Bath-houses in Romano-British Villas*, M Phil Thesis, Univ of London Inst of Archaeol.

Ruesch, A, 1976 Die Militärlager und die Zivilsiedlung in Rottweil am Rekam, *Aufsteig und Niedergang der Römische Welt*, **11**, 5.

Salway, P, 1958 Civilians in the Roman frontier region, *AA (4)*, **36**, 227-44.

Salway, P, 1965 *The Frontier People of Roman Britain.*

von Schaurbein, S, 1974 *Die römischen Militäranlagen bei Halbern: Bericht über die Forchungen Seit 1899*, Munster.

Schmid, W, 1932 Die römische Postatabian Noreia in Einöd, *Jahreshefte des Osterreichische Archaölogische Instituts*, **27**.

Sellye, I, 1939 Les bronzes émaille's de la Pannonie romaine, *Dissertationes Pannonicae*, Ser 2, fasc 8.

Shackley, M L, 1975 *Archaeological Sediments: A Study of Analytical Methods.*

Shaw, R C, 1926 Excavations at Willowford, *Trans Cumberland Westmorland Antiq Archaeol Soc*, **26**, 429-50.

Sheldon, H, Excavations at Toppings and Sun Wharves, Southwark 1970-71, *Trans London Middlesex Archaeol Soc*, **25**

Simpson, G, 1962 Caerleon and Roman forts in Wales in the Second Century AD: pt 1 - Caerleon and Northern Wales, *Archaeol Cambrensis*, **111**, 103-66.

Simpson, G, 1963 Caerleon and Roman forts in Wales in the Second Century AD: pt 2 - Southern Wales, *Archaeol Cambrensis*, **112**, 13-76.

Simpson, G, & Rogers, G, 1969 Cinnamus de Lezoux et quelques potiers contemporains, *Gallia*, **27**, 3-14.

Smythe, J, 1940 The Roman pigs of lead from Brough, *Trans Newcomen Soc*, **20**.

Stanfield, J A, 1929 Unusual forms of terra sigillata, *Archaeol J*, **86**, 113-51.

Stanfield, J A, & Simpson, G, 1958 *Central Gaulish Potters.*

Steer, K A, 1960-1 Excavations at Mumrills Roman fort, 1958-60, *Proc Soc Antiq Scot*, **94**, 86-132.

Stead, I, M, 1980 Rudston Roman Villa, *Yorkshire Archaeol J.*

Sutherland, C H V, 1939 *The Romans in Spain.*

Tassinari, S, 1975 La vaisselle de bronze, Romaine et Provinciale au Musée des Antiquités Nationales, *Gallia Suppl*, **29**, Paris.

Taylor, H, 1883 *Historic Notices of Flint.*

Taylor, M V, 1922 Roman Flintshire, *J Flintshire Hist Soc*, **9**, 58-96.

Thomas, C, (ed) 1966 *Rural Settlement in Roman Britain*, CBA Res Rept, **7**.

Thomas, K D, 1974 in Sheldon, H, 1974

Thomas, T M, 1961 *The Mineral Wealth of Wales and its Exploitation.*

Thompson, F H, 1965 *Roman Cheshire.*

Thompson Watkin, W, 1884 *Roman Cheshire.*

Toller, H, 1977 *Roman Lead Coffins and Ossuaria in Britain*, BAR, **38**.

Toynbee, J M C, 1971 *Death and Burial in the Roman World.*

Toynbee, J M C, 1972 Roman Dover: The painted Rooms, *Kent Archaeol Review*, **29**, 262-4.

Toynbee, J M C, 1973 *Animals in Roman Life and Art.*

Trotter, M, & Gleser, G C, 1958 A re-evaluation of stature based on measurements of stature taken during life and of long bones after death, *American J Phys Anthrop*, **16**, 79-123.

Tylecote, R F, 1962 *Metallurgy in Archaeology.*

Tylecote, R F, 1970 The composition of metal artefacts: a guide to provenance? *Antiquity*, **44**, 19-25.

Wacher, J, 1960 Catterick Bridge, *J Roman Stud*, **50**.

Wacher, J, 1974 Towns of Roman Britain.

Wallace, C, & Webster, P V, 1987 Black-burnished jugs and lids, *J Roman Pottery Studies*, **2**.

Walthew, C V, 1974 The town house and the villa house in Roman Britain, *Britannia*, **vi**, 189-205.

Weaver, P, 1965 Freedmen procurators in the Imperial administration, *Historia*, **14**.

Webster, G, 1953 The lead mining industry in North Wales in Roman times, *J Flintshire Hist Soc*, **13**, 5-33.

Webster, G A, 1958 The Roman military advance under Ostorius Scapula, *Archaeol J*, **115**, 49-98.

Webster, G, 1969 *The Roman Imperial Army.*

Webster, G, 1976 *Romano-British Coarse Pottery: A Student's Guide*, CBA Res Rept, **6**.

Webster, G, & Charlesworth, D, 1958 *The Roman Site at Wall, Staffs.* HMSO.

Webster, J, 1975 Objects of iron, in Cunliffe, B W, 1975, 233-47.

Webster, J, forthcoming a, The small finds, in Williams, J, and Hinchliffe, J, forthcoming.

Webster, J, forthcoming b, The bronze small finds, in Manning, W H, forthcoming.

Webster, P,V, 1971 Melandra Castle Roman Fort: Excavations in the civil settlement, 1966-9, *Derbyshire Archaeol J*, **91**, 59-118.

Webster, P V, 1976 Severn Valley Ware: a preliminary study, *Trans Bristol Gloucs Archaeol Soc*, **94**, 18-46.

Webster, P V, 1982 Romano-British coarse pottery in North-west England - an introduction, *Lancs Archaeol J*, **2**, 13-31.

Weetman, M, 1986 Trial Excavations at the Croes Atti Lane Roman Site, Flint, 1986, *Archaeology in Wales*, **26**, 15-16.

Welker, E, 1974 *Die Römischen Gläser aus Nida-Heddernheim*, Frankfurt.

Wenham, L P, 1968 *The Romano-British Cemetery at Trentholme Drive, York.* H M S O.

Wheeler, R E M, 1923 Segontium and the Roman occupation of Wales, *Y Cymmrodor,* **33**.

Wheeler, R E M, 1925 *Prehistoric and Roman Wales.*

Wheeler, R E M, 1926 *The Roman Fort near Brecon.*

Wheeler, R E M, 1930 *London in Roman Times,* London Museum Catalogue No 3.

Wheeler, R E M, & Wheeler, T V, 1928 The Roman Amphitheatre at Caerleon, *Archaeologia,* **78**, 111-218

Whiting, W, et al, 1931 *Report on the excavation of the Roman cemetery at Ospringe, Kent.* Res Rept Soc Antiq of London, **8**.

Whittick, G C, 1931 Notes on some Romano-British pigs of lead, *J Roman Stud,* **21**, 256-64.

Whittick, G C, 1932 Roman mining in Britain, *Trans Newcomen Soc,* **12**.

Wightman, E M, 1970 *Roman Trier and the Treveri.*

Willhite, L C, 1976 *The praetorium: a study of 34 selected praetoria in Britain and on the German frontier,* MA thesis, Univ of London Inst of Archaeol.

Williams, D F, 1977 Black-burnished ware from Mumrills: a re-appraisal of sources by heavy mineral analysis, in Dore, J, & Greene, K, (eds) 1977, 177-88.

Williams, J H, & Hinchcliffe, J, forthcoming *Excavations at Wilderspool 1966-69 and 1976.*

Wilson, D R, 1974 Roman Britain in 1973, *Britannia,* **v**, 413-4.

Young, C J, 1977 *Oxfordshire Roman Pottery,* BAR, **43**.

Zienkiewicz, J D, 1986 *The legionary Fortress Baths at Caerleon : II The Finds.*

Zeuner, F E, 1963 *A History of Domesticated Animals.*

www.ingramcontent.com/pod-product-compliance
Lightning Source LLC
LaVergne TN
LVHW071218210926
848482LV00015B/847